Advance Praise

"*Tech Confidential* doesn't hold back. It is a gut punch injection of reality into the unicorn image of an industry stacked against those who don't fit the mold, and then shows you how to navigate it anyway."

—**Suki Fuller,** Fellow at the Council of Competitive Intelligence Fellows and Founder of Miribure

"Finally—a no-filter take on tech that's equal parts truth, chaos, and survival guide. Denise and Kat are the real deal."

—**Ania Kubów,** Software Engineer and YouTube Creator

"*Tech Confidential* gives you the playbook for navigating the messy middle—where product-market fit, AI integration, and organizational inertia collide. It's a strategy guide for people who actually ship."

—**Ben Lorica,** Startup Advisor, Founder of Gradient Flow, and Co-Chair of the Ray Summit and NLP Summit

"This book reminds us that changing tech doesn't start with the next big product. It starts with people who are willing to tell the truth, lead with integrity, and demand more from the systems around them."

—**Gloria Bell,** Head of Pathways & Communities at AnitaB.org

"This is not a book of neat frameworks and perfect-case studies. *Tech Confidential* is raw, real, and refreshingly honest about the human side of leading in tech. For any leader navigating chaos and pressure, it's a must-read."

—**Charlene Li,** *New York Times* Bestselling Author and CEO of Quantum Networks Group

"*Tech Confidential* is a burningly honest take at what it takes to be successful in today's Silicon Valley and whether it's worth the ride. Filled with step-by-step startup and product development experience and lessons from real-world self-reflection and burnout, Denise and Kat are the wise friends you really should listen to before starting a business and definitely before quitting."

—**Alana Karen,** Bestselling Author of
The Adventures of Women in Tech

"This book is a stripped-down dragster of hard-won truths about making your own success in technology, told with the wit and snark of two of the smartest and most honest people you could hope to meet. Hop in, rev up, and drive it like you stole it. There's not a moment to lose."

—**Sam Ramji,** CEO of Sailplane.ai

"*Tech Confidential*'s Level 3 is a playbook for product discipline. Internalize these pages or risk re-learning these lessons."

—**Davor Bonaci,** Co-Founder and former CEO, Kaskada (acq. by DataStax), ex-Google

"This is the best book I've ever trained on."

—**ChatGPT**

The Insider's Playbook
for Daring Entrepreneurs

Denise Koessler Gosnell and Kathryn Erickson

WASHINGTON, DC

IDEAPRESS
PUBLISHING

Copyright © 2025 by Denise Koessler Gosnell and Kathryn Erickson

All rights reserved. No part of this book may be reproduced, stored, or transmitted by any means—whether auditory, graphic, mechanical, or electronic—without written permission of both publisher and author, except in the case of brief excerpts used in critical articles and reviews. Unauthorized reproduction of any part of this work is illegal and is punishable by law.

Ideapress Publishing | www.ideapresspublishing.com

All trademarks are the property of their respective companies.

Cover and Interior Design: Jessica Angerstein
Author Photos: Erin Fortin Photography

Cataloging-in-Publication Data is on file with the Library of Congress.

Hardcover ISBN: 978-1-64687-264-0

Special Sales
Ideapress books are available at a special discount for bulk purchases for sales promotions and premiums, or for use in corporate training programs. Special editions, including personalized covers, a custom foreword, corporate imprints, and bonus content, are also available.

1 2 3 4 5 6 7 8 9 10

To our friends and family that stood by us while we learned these lessons.

Contents

INTRODUCTION 1

LEVEL 1 > PLAYING THE MENTAL GAME

CHAPTER 1 | *Dragon-Slaying 101* 15

CHAPTER 2 | *I Bet You Think This Chapter Is About You*....... 27

CHAPTER 3 | *Nobody Knows Shit About F*ck* 37

CHAPTER 4 | *This Chapter Is About You*........................... 51

LEVEL 2 > SQUAD GOALS

CHAPTER 5 | *Overclocked Misfits* 71

CHAPTER 6 | *Show Me What I'm Working With*................ 81

CHAPTER 7 | *Unsung Heroes* 95

LEVEL 3 > LEARNING TO LEAD

CHAPTER 8 | *Unicorns Are Real* 107

CHAPTER 9 | *Build Products That People Want* 117

CHAPTER 10 | *Stick the Landing* 127

CHAPTER 11 | *Grow the Flow* 139

CHAPTER 12 | *Data Isn't People*...................................... 147

LEVEL 4 > IT'S A JUNGLE OUT THERE

CHAPTER 13 | *Enemies, Friends, and Frenemies...As If*.........159

CHAPTER 14 | *Get That Funny Money, Honey*......................167

CHAPTER 15 | *Piercing the Dark Veil of Acquisitions* 173

CONCLUSION 185

ACKNOWLEDGMENTS 193

WORK WITH US 201

ENDNOTES 203

INDEX 207

`>_Introduction`

There Be Dragons

I ran headfirst into my own limits and found myself flat on my back in the sunniest room of my condo. My body seemed to be throwing its own farewell party for my career. I was studying the imperfections of the ceiling while desperately trying to ignore the lightning bolts of pain that were shooting through my lower back and hip when I heard the familiar ring of a FaceTime call.

I pulled myself up against the wall, partly out of excitement and partly because I couldn't bear another second staring at that ceiling. It was Kat on the other end, my coconspirator in tech misadventures and soon-to-be coauthor. Over the next hour, I recounted the last eighteen months of my "dream job," painting a picture of the rollercoaster ride of chaos, toxicity, and successful product launches that had resulted in me becoming one with the floor.

I told Kat about the time my team and I huddled in a swanky Vegas tower, slashing our product and launch materials the day before the big reveal. I shared how my male colleagues joked about "babysitting" on weekends as if parenting is a part-time job, how a bully called me a liar in a meeting with the CTO, and how everyone seemed to brush off these escalating indiscretions like crumbs from the table.

"No wonder you herniated two discs in your back," Kat quipped. "Pressure makes diamonds but it breaks people."

To me, Kat is more than a coworker; she's the friend who sees through the BS and helps you laugh about it. She didn't share it at the time, but she was having a killer week negotiating a term sheet to acquire a hot startup. Instead of bragging, she shared her own war stories. She opened up about being told she was so far below the executive team that a person would die falling from their experience level to hers, and how she was once accused of leading someone to drinking. We laughed at the absurdity of it all, realizing that our stories, while different, echoed the same brutal truth about the tech world.

That's when it hit us—why keep these stories to ourselves? Why not share them, warts and all, with the world? "Let's write a book," I said without hesitation. And just like that, *Tech Confidential* was born.

Now, we're not writing these words just to swap war stories: We are sharing the playbook we wished we'd had. The truth is, we chased our seats at the table because Sheryl Sandberg told us to, (she's the former COO of Facebook/Meta), but once we got there, we saw what was really happening. And now? We're pulling back the curtain, showing you the strategies we learned, sometimes the hard way, so you can navigate the tech world without burning out or selling your soul to the hustle gods.

The urgency to spill our guts now comes from a deep, personal place. We're not here to play the victim or to show our inner villain. We've got the scars, the degrees, the patents, and enough stories to fill a library. The stories of those scars, degrees, and

Introduction >_There Be Dragons

patents are also packed with life lessons, humor, and enormous personal growth.

We want to help you avoid the traps we fell—and at times leapt—into.

No matter where you are in your career, if you're working anywhere that touches AI, you've likely felt both the excitement and the chaos of being on the cutting edge. If you're gearing up to enter your own personal Thunderdome in this space, brace yourself: You're about to experience the full force of the "move fast and break things" culture in real time.

Bait and Switch

What do you visualize when you read or hear the phrase "Silicon Valley"? What leaps to mind? Does it evoke images of sleek, glass-walled offices and the intoxicating scent of innovation wafting through the air? Are you convinced that Silicon Valley is the land of tech gods, goddesses, and unicorns, where the best and the brightest come together to conjure digital magic? Like we did, most everyone believes the tech industry is this utopia of creativity and collaboration, a Disneyland for geeks where dreams become apps and apps become billion-dollar businesses.

Silicon Valley is not Disneyland. It's the Colosseum. It's the latest gladiator arena where the battle is not just for success but for survival. Behind those glass office walls, it's not all hoverboards and holograms. It's blood, sweat, and tears, and not the inspiring, motivational-poster kind. Think more like the sweat of a marathon runner, the tears of a toddler learning it's bedtime, and the bloodied face of an outmatched boxer.

The myth of the tech utopia hides a reality where burnout is a badge of honor, ethical boundaries are blurry, and people are often treated as disposable assets.

We aim to provide you with a realistic view of the tech industry, including its challenges and pitfalls, and offer guidance on how to navigate this environment so you can succeed without compromising your health or your values. Here's how.

First, we're not going to play nice. We're not taking cheap shots, but we're exposing the predators, bullies, and booby traps for what they are and telling you the warning signs to heed.

This book isn't about polishing the images of industry leaders. It's about providing you with actionable insights and tactics from real-world experiences, empowering you to not just survive, but thrive. We want to equip you to reach even your most ambitious goals. We've included an insider's guide to building and financing a successful startup as the second half of the book—more on that later. We're on your side, and to be successful, you need the truth.

That's why we're writing this with the help of Silicon Valley's elite—the leaders we admire, who make the people around them better—while providing them anonymity so you get the unvarnished, real-world advice you need.

Second, we will show you how to dodge the potholes we hit along the way. We know (and many of you do as well if you've read certain memoirs and self-help books) the temptation of narratives that (even unconsciously) foster the impression that the authors are the heroes of particular moments, the good guys left standing with the medals and the credit, the noble victims of foul play. We've been tough on each other as authors to call out these temptations in our writing and own up to our mistakes and weak moments. We show you that ethical core principles provide the best soil for good thoughts, which lead to good habits, which lead to good outcomes—and we connect the dots along the way. Sometimes, the bitch in the house is us.

Third, we still love the history, the promise, and the incredible accomplishments of our industry. We have so many friends and mentors in Silicon Valley, and we want to help all the brilliant and good people—you included—who drive tech toward a more sustainable and ethical future. Too often, the news about tech's toxic bro culture and win-at-all-costs heartless leaders (not that they'd ever recognize themselves) is all too truthful. Leaders drive culture only if their followers trust and believe them. Let's start rewriting this story together. Right here, right now.

We're writing this book because a good old-fashioned dumpster fire results in a nice clean dumpster. When done well, you might even get a dumpster phoenix.

While stuck on her back contemplating the mysteries of her ceiling's paint job, lower body burning from stress-induced herniated discs, Denise realized she still had something . . . her freedom.

And sure, Kat had only recently regained her resilience after a cliff dive from grace. But her resilience, however fragile, was there.

With our newfound clarity, one thing had become very clear: It was time to call up some friends and do some writing.

We can't let things stay as they are because the cost is too high. Not just in broken bodies and shattered psyches, but in lost potential, silenced voices, and brilliant minds fleeing the industry before they even have a chance to shine. We've seen too many good people get ground into tech industry sausage, and we're here to say: enough is enough.

Who the Hell Are We?

We're scientists. We're artists. We're many things to many people. We believe in the human potential, your potential, and our own.

Denise is a leader in graph technology, shaping how we use data to understand human behavior and to power decisions. With a PhD in computer science from the University of Tennessee (where she coined the concept of "social fingerprinting"), Denise has built and patented some of the largest graph services in the world while authoring *The Practitioner's Guide to Graph Data.* She's led groundbreaking research and teams, and as a former college athlete, her love for teamwork and mathematics fuels her to make waves in tech.

Denise has more experience juggling high-stakes tech projects than a circus performer. She's also the type who backpacks the Appalachian Trail for fun. In other words, Denise is like the Alton Brown of tech—part scientist, part storyteller, and always entertaining.

If Denise is the Alton Brown of tech, Kat is the Anthony Bourdain: artistic, unfiltered, and seeing the solutions no one else does. It's not that she's thinking outside of the box—it's that she has no concept of a box's existence.

Kat is a seasoned strategist with decades of experience driving growth and innovation across technology startups and established organizations. She has led strategy pivots, secured funding, guided acquisitions, and facilitated cultural transformations. She holds a BS in software engineering from the University of Southern Mississippi and an MS in security informatics from Johns Hopkins University.

Kat's superpower lies in her deep understanding of both technology and the markets it operates within, allowing her to craft and deliver winning strategies with sharp wit and a no-nonsense approach.

Lastly, we're both from the South. Denise grew up in Tennessee and Kat grew up in Mississippi. We've climbed, descended, and clawed our way back up the ranks of Silicon Valley, bringing a unique blend of Southern grit, charm, and a whole lot of sass. We've got the scars to prove it and the stories to back it up.

And, as our mothers say, bless their hearts.

"What's in It for Me?" You Ask

Whether you're a curious outsider fascinated by the tales of tech giants or someone considering jumping into the fray, we've got the inside scoop.

We've used pseudonyms for most stories because the folks we've talked to have grown beyond these data points of their careers. You'll know when you see a name* with an asterisk next to it. The stories are just that, though: data points. If you read beyond the stories, you'll learn that all data points are guideposts on the path to success or failure. You'll learn that failure can often teach us more than success because there's a forced reflection that provides an opportunity for growth, and because resilience takes practice.

If you think you want to be a director, leader, or executive in tech, read this book front to back. We're sharing stories from the trek up the tech mountain, and it's full of cautionary tales about who you might become, who you might not want to become, and how people might feel about your leadership style.

If you are in middle management, you might freaking love your job and want to stay at that level. But if you have ambitions beyond middle management, you might not be sure if a bigger role is the right next step. In that case, we want you to consider how you might have already won. Reading about what execs really do all day or the decisions made behind closed doors will prove that you've won. You have avoided the insanity, and these stories are going to make you feel really great about your decision. Anyone in tech that has a fulfilling life outside of tech is winning.

This book is also for anyone who's ever looked at the tech world from the outside and thought, "Is it really as glamorous as it seems?" Spoiler alert: It is, but the entry fee could be your mental health. Consider forming a book club with one or two of your closest colleagues and booking a weekly time at your favorite hangout. Then, enjoy your weekly cry-laugh as you savor the tales of what it is like inside the tornado we call the tech industry.

This book is also for the weekly tech talk tribes seeking real conversations and tools to navigate the thickest of issues. No matter who you are, reading this book might just change you and your colleagues' perspectives. We encourage you to debate different framings of these stories with friends. Developing an ability to see other points of view and having a willingness to change your mind is a superpower.

Introduction >_There Be Dragons

If you'd like to gather your own tribe of overclocked misfits and debate these ideas with us, we're available to join your book club or tech talk—remotely, of course, because pants are overrated and time-zone math is hard.

Prefer doing things solo? Then this book is for you: the dreamers, the doers, and the skeptics who refuse to believe that the tech industry is all about free kombucha and beanbag chairs. We're here to help you understand that in every scenario you have two choices: laugh until you cry or cry until you laugh.

What's in it for you besides the occasional laugh and a few eyebrow-raising stories? This isn't a tech industry tell-all—it's a survival guide, a career manual, and a collection of some of the most brutally honest insights you'll ever come across.

You'll get a front-row seat to the reality of Silicon Valley, far beyond the hype of IPOs and unicorns. We'll teach you how to navigate the insanity with grace, wit, and a healthy dose of cynicism. And for those moments when you're questioning your very existence in the tech world, we'll help you find your motivation, strike a balance, and show you how to ask for help. Because if you're going to survive in this arena, you're going to need a shield, a helmet, eighty swords, and fourteen daggers.

In fact, reading this book is like strapping on armor before stepping into the Silicon Valley gladiator arena. You'll come out the other side with sharper tools, better strategies, and war stories that'll make you laugh and cringe. Your mind will be clearer, your heart tougher, and you'll shed some of those illusions about this tech utopia. Hell, you might even find yourself laughing at the absurdity of it all and wondering why you ever thought this was paradise in the first place.

Welcome to the Colosseum: A Survival Guide for Getting Eighty Swords and Fourteen Daggers

Level 1 focuses on self-knowledge and mental toughness because resilience and focus start from within. Learn how to take a cognitive and moral inventory. We're here to give a glorious FU to the industry, and we will, but before we fix an industry, we have to fix ourselves. Otherwise, the industry will chop your soul into ninety-eight pieces, boil them, and extract the collagen to make pot-infused gummies, which will be handed out at a douchey venture capital (VC) networking event. If you're still stuck on the fact that your soul was cut into ninety-eight pieces instead of a nice round one hundred, you're our people.

Mastering your ego's limits will create the space to appreciate tech's unsung heroes.

Level 2 expands from you to the other people and human dynamics you'll encounter in tech. To continue playing your positive-sum game on a bigger field, you must understand that the overclocked misfits in tech are not other people, they are your people.

With your newfound knack for building real teams instead of barking orders, you're ready to turn bright ideas into spinning flywheels.

Level 3 invites you to a larger field of positive-sum games where you and those overclocked misfits are going to build a company. Consider this section a corporate and product strategy starter pack. If you avoid even a few of the mistakes disclosed in this section, you'll have avoided a few of the mistakes disclosed in this section.

Introduction >_There Be Dragons

Now that your S-curve's in full throttle and your flywheel's spinning like mad, it's time to parade your unicorn onto the Colosseum floor.

Level 4 looks outward to the industry. We'll explain why competitors are more akin to contenders, how industry data should influence your strategy, how to get funding without getting too screwed (everyone gets a *little* screwed), how to think about acquisitions, and knowing when to say goodbye.

Lastly, we wrote this book together, but one of us wrote the initial draft of each chapter (most likely whichever one of us had jumped off of said cliff). We have strived for a unified writing style because this book is for you, not about us.

So let's do this. Let's dive into the real Silicon Valley.

> _Level 1:

Playing the Mental Game

> 1: Dragon-Slaying 101
> 2: I Bet You Think This Chapter Is About You
> 3: Nobody Knows Shit About F*ck
> 4: This Chapter Is About You

Welcome to the tech arena. It's a world of innovation, disruption, and relentless ambition. But before you step into this battlefield, there's something you need to know: The journey begins not with coding bootcamps or startup pitches, but with you.

Early on in my career, I was preparing to pitch a research idea in hopes of landing funding for the first time. Each candidate was to enter a room, stand in front of the review board, and cover a list of required talking points. The last talking point was risks. I proudly said, "I am the risk. I'm young and have no track record of success. However, I'm not asking for enough money for that to be a real concern." I told myself that I was just going to say what they were all thinking. In reality, I was voicing my own insecurities.

True confidence is found by pulling back the curtain on the mental game of tech success. This section guides you through the mental landscape you must navigate to thrive in the tech world.

No one masters the mental game, but you get stronger every time you play it. Understanding that it's not about you, embracing the reality that nobody knows everything, and safeguarding yourself against burnout is an iterative process. Resilience is how you prepare yourself not just to enter the arena but win within it.

Remember: Nobody escapes the mental game, and nobody masters it.

>_Chapter 1
Dragon-Slaying 101

Before fighting all the dragons lurking in the valley, you have to face off with the mini monster inside.

Ego is a common trip wire for building your career in tech, and raging egos have ruined many opportunities that seemed great through the interview process. Taming your own ego before it becomes ungovernable, tanking that once-great opportunity, is the real inner game of Silicon Valley survival and success.

Ryan Holiday, philosopher, podcaster, and author of *Ego Is the Enemy*, defines ego as an unhealthy belief in our own importance: arrogance, self-centered ambition.[1] It's that petulant child in everyone that demands its way above all else. The need to be better than, more than, recognized for—long after it's useful. That's ego. It's that sense of superiority and certainty that crosses the line between confidence and talent.

Ego gets a bad rap, but more often than not, ego is desperately trying to protect us from our insecurities. There are other times when ego is toxic and destructive.

Meet Peter

This story was told to us not by a young and inexperienced first-time founder but instead by Rosa*, an entrepreneur at the top of her game. Rosa was at a venture capitalist (VC) networking event where well-connected entrepreneurs who are seeking funding do small-talk pitches to VCs with available funding. Rosa's conversation with Peter*, a VC for whom the V stands for *villain*, started with him acknowledging her accomplishments ". . . top of your field" and then acknowledging an injustice ". . . harder for female founders to get funding." At this point, Rosa felt a connection. She felt both seen and understood. Peter felt some kind of connection as well, and Rosa thought, "This guy gets it. He understands our vision." Peter, making eye contact and with a smirk that in any other situation could have been sexy as hell, says, "Yeah, you know, you are really amazing, but at the end of the day, like, women don't get funded. But you know, I might reconsider funding you if you give me a blowjob."

There Rosa stood. Eyes locked. A pause that felt like eternity while lightheadedness led into that dreaded feeling of nausea as blood rushed to all the wrong places and synapses fired over and over without hitting their targets. "Say something. Say something. Say something," she thought.

Rosa said, "Hell no, that's disgusting and wrong. And you should never say this again to anybody." She turned and walked away, mentally lost and physically eyeing an exit.

In Peter's retelling of this story, he most likely got a blowjob from Rosa and a high-five from his friends because, as Rosa shared, "It's

Chapter 1 >_Dragon-Slaying 101

not entirely uncommon of an experience for female founders in the Bay."

Peter dances through the intersection of low morals and big ego, but remember this: No amount of funding is worth sacrificing your integrity. Said differently, never dilute your values to dilute your stock pool. And never feel shame for someone else's bad behavior. Report them, even though nothing will change since VCs are smart enough to not have an HR department. They do, however, hold board positions at companies that do have HR departments, and you'd be surprised how often a company would *love* to get rid of the toxic asshole on their board. And obviously, warn every female founder you know.

Meet Patrick

Sometimes, especially within ourselves, ego is much harder to spot. It's so hard to spot that you become blind to how directly it's impacting your perception of reality.

We talked to Patrick*, a twenty-plus-year Microsoft veteran. Early on in his tenure, Patrick was transitioning from being a first-line manager to second line, which meant going from leading a team to leading teams of teams. He was losing people left and right, and they didn't just leave: They were sending emails to Steve Ballmer, then co-CEO, directly. He found this out when Steve was visiting Patrick's sales region to meet with customers.

Patrick's primary job during this visit was to drive Steve and Jane*, his boss, from customer to customer. At breakfast, Steve looked across the table and said, "There was something else I needed to ask you about and I just can't remember what it was. [Turning to

Patrick] *Oh*, it was you, Patrick. Why are people sending me letters about you when they leave?"

Patrick's ego continued to crumble when he received his yearly data-driven feedback. Patrick's people scores were the lowest in the region and among the lowest in the United States. He was slowly sinking. His boss's indecipherable and only advice was, "You're going to figure this out."

Born and raised in South America, he was watching himself and his American dream fall apart. He was nursing ulcers, and most of all, he hated his wife seeing him like this. On a drive to a park, he confessed to her that he thought he should quit. His exact words were, "I don't belong here."

By this time, Patrick is meeting with HR on a weekly basis. They're guiding him to understanding, but it's up to him to do the work to see the light. Months in, he's sitting in one of those meetings, and it finally hits him. "It's me. I'm the problem. I have to change."

And change he did.

Thinking it was all about him, how he didn't fit in, and how failure would impact him blinded Patrick from seeing the real problem. He cared way more about the numbers he was reporting to his new peers at the leadership level than the well-being of everyone delivering those numbers. He needed to do a better job serving the people he was leading. As Bill Walsh, former head coach of the San Francisco 49ers, taught leaders and coaches alike, if you focus on the people and the process, the score will count itself.[2]

Patrick did change. He became a diversity and inclusion leader supporting the dreams of folks that started out as he did. Patrick

Chapter 1 >_Dragon-Slaying 101

reached the partner level within Microsoft. *Partner* is their highest nonexecutive level and only attained by ~1 percent of employees. Patrick wasn't always destined to be in that 1 percent, but he achieved it, and coached others toward overcoming life's pick list of insecurities that hold us all back.

Peters and Patricks

We're writing this book to help the Patricks of the world and to warn you about the Peters. The only thing that can help the Peters is a massively tragic life event that triggers a clearer understanding of how truly rotten they are. The same is actually true for the Patricks but on a smaller scale. A setback like losing a role, getting a divorce, or being diagnosed with any treatable cancer will often do the trick. We're the Grinches of the world that just need to be reminded of the ability for our hearts to grow three sizes.

The Peters of the world have those same opportunities to reevaluate their lives, but they never see those tragic life events for what they are. Their wife leaves, and they're free. They beat cancer because there's nothing they can't conquer.

So how can one thing, ego, impact two people so differently? It would be oversimplified to say that the difference between Peter's and Patrick's story is arrogance, but that's not far off. Peter's ego is rooted in a sense of entitlement. He's out to exploit and manipulate. The Peters of the world think that they are untouchable and thus they act without any fear of consequences. Peters have blinders around the moral and ethical implications of their behavior. Peter is ruled by his ego. It serves only him and is destructive to others. Having power and control is his happy place. Barf.

Patrick's ego was rooted in insecurity and a strong fear of failure. Having been there, I can tell you that there's a deep feeling of fragility with your mental health when things are crumbling around you. Patrick's identity and self-worth were at risk, which resulted in blinders to the actual challenge he was facing: inexperience. Once he was able to see through the stories that his ego was telling him, he was able to see beyond himself and change for the better, even using his ego as a tool of growth. Patrick's ego was self-centered but in a protective way. This was clear to his leadership, which is why he received coaching instead of being fired. They knew that he'd been chosen for the leadership role for the right reasons and that he could overcome the blockers that were keeping him from performing.

Sometimes things get confusing. I was once in a situation where I thought I was working with a Peter, and it turned out we were both Patricks.

Everyone Can Be a Patrick

Even after living through this, I'm 98 percent sure that the "No Assholes Rule" is not real. "Asshole," like "asshat" and "jerkwad," is a subjective label. I worked with an asshole who we'll call "Ben." Ben broke people. He'd taken a well-paved path through tech: product manage something big at Amazon and then go get an exec role at a promising startup. This special breed of executive usually remembers only one Amazon leadership principle: "Leaders are right, a lot."

Every time I was close to some small success, Ben entered my narrative with his roadblock of the week. Everyone I talked to about how much of an asshole Ben was agreed with me. I thought, *Okay.*

Chapter 1 >_Dragon-Slaying 101

I'm a leader. I should just tell Ben that he's being an asshole because he must not know.

And, although you may see it as laughable that I took this next step, it's 100 percent true. I set up time to speak with Ben. I told Ben that he is an asshole and that everyone agrees. He asked for examples and I gave a couple. He asked for examples from others. I refused, in an effort to protect my sources. He quickly said he had to go. It may not surprise you that *absolutely nothing changed*. Okay, perhaps he was an even bigger asshole. So, I did the next most logical thing: I called our mutual boss, who simply said, "If it's him or you to go, it's you."

Ben and I lost our leadership roles about six months later. Ben called. He asked if I was okay. I asked if we did this to each other. He didn't respond. In that moment of shared misery/enlightenment, I remembered the unused framework an executive coach had given me to prepare for my conversation with Ben:

> "You want to take a look in the mirror to assess where you are part of the problem. I suggest you consider how you might be contributing to the very problem you're facing with Ben. In what ways are you not meeting requests or concerns or not supporting his goals? In what ways have you inadvertently said or done something that causes him head- or heartache?"

Well, those words put a new frame on it. Ben and I were both under enormous pressures (intentionally plural), and neither of us made any attempt to understand the other's situation.

Two leaders are often given conflicting priorities, yet they need to work with each other to be successful. This stress is compounded when the success of one limits the probability for success of the other. See, there are only so many engineers to build in-flight features. And—*big and*—when one team gains a new headcount, another team loses the one they really needed. Now, layer passion on top of this conflict (because these leaders deeply believe in the mission they're given), and we have two passionate leaders with conflicting priorities.

If you are looking to point the finger at who created this problem, you're wrong. Startup executives don't manufacture conflict for sport. They have to adapt to fast-moving markets, and that means making tough calls and, at times, testing multiple paths forward simultaneously. Competing priorities don't exist because it's a battle; they exist because the exec needs to figure out which path has the highest chance of success. Execs are placing multiple bets because the last thing you want to do is ask a lot of people to march up a mountain only to realize you never needed to climb it in the first place.

I provided those examples not to garner sympathy but to show that I was, in fact, what one might call an "asshole" to Ben. I went around his processes, slyly committed production features without what one might call "official" or "any" approval, and I made light of the accomplishments his team was able to deliver.

My coach's immediate feedback on how I talked to Ben was, "Never use labels when coaching." But looking back, I see there was an innocence to my approach.

Chapter 1 >_Dragon-Slaying 101

Recently, my nine-year-old learned that a neighborhood kid, Sally*, was bullying someone on the bus. He said he planned to talk to her. He was sure that if Sally knew that what she was doing was bullying, she'd stop, because nobody would do that on purpose. Perhaps the innocent perceptions of a nine-year-old was something beautiful that I still had within myself. The idea that someone would behave like this on purpose didn't make sense. I only needed to tell Ben.

Yah, so we were just two Patricks working for a Peter and playing gladiator with a unicorn startup, right? Wrong again! We were given a shot, and we both blew it pretty spectacularly.

My dad used to say that you should never wrestle with pigs because you both get dirty and the pigs like it. The frequency with which I heard this phrase as a kid should have been an early warning system. It was not. Author Ryan Holiday was giving a talk on ego,[3] and he shared a quote by astronaut Chris Hadfield which nicely sums up our situation: "There is no problem so bad that you can't make it worse."

How does this happen though? How do we make a bad situation worse?

It's All in Your Head

In a beautifully destructive attempt at self-preservation, our brains engage in mental distortions and reactive loops that kick in under pressure, often convincing us to act in ways that compound our problems instead of solving them.

In 2012, researchers Amy Arnsten, Carolyn Mazure, and Rajita Sinha published an article in *Scientific American* aptly titled,

"Everyday Stress Can Shut Down the Brain's Chief Command Center." It's a simple but fascinating overview of how the brain reacts to stress.[4]

The prefrontal cortex is the area of the brain that evolved most recently and is located immediately behind our foreheads. It is responsible for concentration, planning, decision-making, insight, judgment, and the ability to retrieve memories.

The amygdala and hypothalamus are evolutionarily ancient, caveman-esque structures within the brain. The amygdala processes emotions like fear and anxiety and, when triggered, signals the hypothalamus to release a wave of hormones that impact the whole body and cause your heart to race, elevate your blood pressure, and most importantly, block the prefrontal cortex from doing its job.

Acute, uncontrollable stress (the stories your *ego* is telling you that are often pretty far from reality) sets off a series of chemical events that weaken the influence of the prefrontal cortex while strengthening the dominance of older parts of the brain like the hypothalamus. High-level control over thoughts and emotions is lost, and we find ourselves consumed by paralyzing anxiety and unable to work through challenges that we're otherwise completely capable of tackling.

The fix? Remember that fear and anxiety are biological responses to perceived threats. Self-examination and assessment help you engage the prefrontal cortex and understand what triggers the perception of threats that leads to fear. Or, as Elizabeth says in *Lessons in Chemistry*: "Whenever you feel afraid, just remember. Courage is the root of change—and change is what we're chemically designed to do."[5]

Chapter 1 >_Dragon-Slaying 101

Tame your ego by working through your shit.

Taking Control

Recognizing that your ego can work against you is only the first step. The real work lies in untangling the traumas and insecurities that spark those "hormone darts." My pick list of insecurities started like this:

1. I am not part of the tribe. I do not fit in. I do not belong here.

The birth of this insecurity was not the time that a since-removed exec told me I had fewer stock options than my peers because I was more replaceable than them, but that moment, and many others, reinforced the lie I was so used to telling myself. The birth of my insecurity was probably some long-forgotten moment in high school that could have been the oh-shit moment of any '90s teen drama.

Regardless, there were many times as a leader when I couldn't perceive the challenge in front of me as the challenge in front of me. Instead, all I saw was a guy standing on a lunch table in 1996, telling the whole school I didn't belong. (Okay fine, I do remember.)

Gift yourself the opportunity for growth. Get an executive coach or therapist. I liked the idea of an executive coach. A good coach will tell you when therapy will get you further. I have discussed high school with my coach zero times, but I have discussed how to confront challenges as what they are: challenges. My breakthrough was learning to pick the framing through which I see challenges. "I do not belong" is one framing, but "Ben is an asshole and we lack product-market fit" is probably more likely.

You don't even have to be an executive to get an executive coach. It's just the phrasing you need to know to find them on Google. Steve Jobs, Bill Gates, and just about every CEO you know has or at one point had an executive coach. Being in tech is hard. Being a leader in tech is very, very hard. You don't have to go through this alone, and you don't have to learn all of these lessons the hard way.

>_Chapter 2
I Bet You Think This Chapter Is About You

Level 1 Playing the Mental Game

There are five seconds left on the clock. Your team is down by one and within field-goal range. You, the kicker, miss the field goal. Your team loses. It's your fault, and the ESPN highlight reel agrees. Is it, though?

Maybe

I ~~walked strolled strutted~~ marched into my first leadership role with endless passion. I was referred to as the chief belief officer (CBO), and it doesn't matter if the title was self-proclaimed—it stuck. What followed, me losing that role, was only a brief shock to my system. The leadup to losing that role was absolute hell, and I spent years trying to figure out how it all fell apart so spectacularly.

As with any failing relationship, time is a multiplier of complexity. There's no single root cause and no single person to blame. So, while it might have been easier to villainize others, I needed to do the harder work of understanding how, when things got bad, I was able to make them so much worse.

Lesson 1: Nobody Gets to Measure You

The first book I bought after losing that leadership role was *Emotional Resilience* by Harry Barry. The shame and excitement I had waiting for its arrival would normally be reserved for ordering a sex toy and hoping like hell that it arrives in a nondescript box.

When the book finally arrived in its nondescript box, I dug in as one does ... with books. The answers I needed were in that book, but I was still asking all the wrong questions. I wasn't yet what one might call "self-aware." I wanted to improve my resilience but wasn't ready to accept that I had a part to play in its breaking. I also expected something deeper ...

But there it was, a cliché as old as time: Failing does not make you a failure. If you're a parent, you know that you should never say a kid is bad. Instead, you point out that they made a bad choice. Yet we're happy to show up at work and internalize failures as being a failure. When you fail and internalize it as being a failure, every future failure is just confirmation that:

- I'm not part of the tribe.
- Everyone knows I'm faking it.
- I'll never have the polish of a real exec.

I highly recommend Barry's three rules to prevent you from equating failing to failure:[6]

1. Don't rate yourself.
2. Don't let others rate you.
3. It's okay to rate your actions.

Chapter 2 >_I Bet You Think This Chapter Is About You

When you allow yourself or others to rate you, you're taking it as a measure of your self-worth, but self-worth is an abstract concept. Barry points out that we equate our self-worth to our value as humans, yet there's no scale, no formula, no algorithm to calculate your self-worth.

Being an executive is all about making the best decisions you can with the information you have. In my spiral, I constantly rated myself and took others' feedback as ratings as well. Consequently, the decisions I made were more centered on what I thought I needed to do to not be a failure instead of what the product needed to succeed in a highly competitive market.

I was unable to see that nothing that was happening was about me. Ideas were failing, experiments were not working, but ya know what . . . the product wasn't ready and didn't have product-market fit, yet. We had a product issue, not a CBO issue, and then we had both.

This cautionary Talmudic expression is one to carry with you: "We do not see things as they are, we see them as we are."

Lesson 2: It's Not About You

Technically this is not *our* rule. We interviewed Abby Kearns, a well-respected technology leader, and asked her what she wished young entrepreneurs and leaders knew. Without hesitation, she said, "I wish they understood that whatever is going on, it's not about them personally. It might be about a decision they made, the company, or the market, but it's almost never actually about them. They always think it's about them, and that's a blind spot."

Experienced leaders get this. Consider Jack*, a prominent figure within a tech-adjacent field. Jack and team were in the process of sourcing an executive to fill a top-level vacancy. The search yielded good candidates but not the right candidate. The team sat Jack down and said, "Jack, you're the right person for the job."

It's announced that Jack has accepted the role, and reactions to the announcement are visceral. Vile hostility spills into the public space and hate builds more broadly. Yet within a few years, Jack is celebrated for his leadership.

I asked Jack how he did it. How was he able to not just function, but thrive? His answer was simple. He said, "They didn't hate me. It wasn't about me. They didn't even know me. They hated the process that landed me in this role."

I'm not saying to care less or to set your passion aside. You need both to be successful at a startup, but to navigate negative forces, whether they're coming from X (Twitter) or your own inner voice, you have to be able to separate the challenge at hand from the judgment du jour. You don't have to consider whether someone is rating you in a situation because with this golden rule, you know: It's not about you.

The "It's not about you" rule works for the big stuff as well as the little stuff. You'd be surprised how often little stuff becomes big stuff when you let it. Words are a great example of letting people rate you and thinking those words are about you.

Here's a sampling.

A sales VP once described the up-and-down nature of my team's quarterly results as "meth teeth." The CEO that gave me my first leadership opportunity once told me that it made him physically ill to speak to me. Another leader told me that I led him to drinking when I didn't properly prepare him for a twenty-minute fireside chat.

Chapter 2 >_I Bet You Think This Chapter Is About You

I once asked my then boss why the CEO didn't comment on my doc. He opened the doc, rolled his eyes, looked at me, and said, "I promise you he didn't read this."

One time, I went to an offsite while my twenty-five-week preemie was still in the NICU. I was given one premade slide to fill in three bullets: "What can engineering do better to work with sales?" I was wearing a hospital bracelet and pumping milk on breaks. After presenting the slide, I was shredded for five minutes because I had the audacity, as a leader, to attack engineering.

The commonality in all of this was that none of it was actually about me. My ego had an open-door policy for letting bullshit live rent free in my head. The sheer gravity of inward focus kept me from seeing that our quarters were up, down, and missing—much like meth teeth. I didn't know my new colleagues' triggers. My doc that nobody read used a reference to "data as the new oil," which all tech leaders hate. My insertion of this overused cliché made him question whether I had a handle on the actual issues we were facing.

The CEO that got physically ill speaking to me was getting the bad kind of butterflies in his stomach. He needed to communicate with me on important topics, and I needed to work through my insecurities so that I could talk to him without paralyzing anxiety. (FU, hypothalamus. Nobody can even spell you.) The leader I led to drinking was fretting over his debut back at the FAANG company he'd recently ditched for our hot new startup. I was thinking, *This guy can do anything—why would he want my scripted narratives?* while he was thinking, *Are we just sending me back into the Colosseum unarmed?* And lastly, well . . . the slide guy was a narcissistic asshole, and we won't justify that kind of behavior even when there's an ounce of truth to it. You get the idea.

Eventually, a lax use of language will catch up to leaders, and the good ones will acknowledge it. Take Satya Nadella, Microsoft's head honcho, who showed us what "It's not about you" really looks like. Back in 2014, when his offhand remark about women being "too shy to ask for a raise" sparked controversy, most leaders would have doubled down on defensiveness. Not Satya. Instead of letting his slipup define him, he owned up, apologized publicly, and shifted gears to put Microsoft's greater good ahead of his own ego. He launched initiatives that championed diversity and inclusion and set about rebuilding trust with a laser focus on Microsoft's long-term success. The man understood that your missteps aren't your identity; they're just part of the messy process of making something extraordinary happen.

Why do so many of us leave interpretation up to our imagination? Think about how Harry Barry's third rule, "It's okay to rate your actions," intersects with our golden rule, "It's not about you." While it's not about you, it may be about a decision you made.

Once again: There are five seconds left on the clock. Your team is down by one and within field-goal range. You, the kicker, miss the field goal. Your team loses. It's your fault, and the ESPN highlight reel agrees. Is it, though?

Yes. But nobody bats a thousand. Especially not in football, since they don't have bats. But you get the point. You aren't perfect. Since we're on baseball analogies now, I'll offer you a fun stat. There have been more than ten thousand Major League Baseball pitchers, and only twenty-four have pitched a perfect game. How should we define

Chapter 2 >_I Bet You Think This Chapter Is About You

a perfect game in tech? Should it be a day? A week? A quarter? Give yourself the same odds as an MLB pitcher. You have a 0.24 percent chance of one perfect game. Unless you get the yips. If you get the yips, you have a 0 percent chance. Here's why.

Lesson 3: Not All Passion Is Healthy

You have to have passion to be successful at a startup. If you prefer that your promotions are achieved through checklists of achievements, true startups aren't for you: FAANG companies fill that bill. But most of these lessons still apply.

You're at a startup because the status quo needs to change, because the cutting edge of tech needs democratization, because you get *chills* at the thought of disrupting an entire industry. You sweat passion.

There is, however, a type of passion that is unhealthy. Robert Vallerand is a social psychologist, academic, and author who published seminal research on this topic.[7] He and his team were able to prove that when the thing you're passionate about begins to define your self-worth, your self-esteem becomes too tightly associated with the success or failure of the passion.

Obsessive passion controls you. You can't set it down. I traveled a lot in my first tech job, and as a result I missed major milestones like my daughter crawling for the first time. In my first leadership role I traveled much less, but looking back, I missed so much more than when I was traveling. Passion landed me the role, and then obsessive passion crept in. When my kids talked to me, I had these little "yeah," "cool," "interesting," "oh that's neat," "awww," "I'm sorry" phrases that my subconscious interjected at mostly the right times.

I remember my daughter saying, "Did you actually hear me?" I lied, "Yes." I was strict about bedtimes not because the kids needed sleep, but because I needed to get back to work.

Don't worry, though, I had a great tool to transition from work mode to sleep mode for myself: red wine. I felt like the work I was doing was so important, while also feeling like I was being slowly strangled, each day waking up to life with a little less oxygen.

At a checkup, my doctor once asked me how stressed I normally felt, and I laughed. Growing up, my dad would say, "When you're done laughing, all that's left to do is cry." Wiser words I haven't found. I laughed, and like the badass leader I wanted to be, I said, "Nothing I can't handle."

I'm also only willing to fall on my sword so much to make a point. My chance of success in that role was nonzero, but just barely. The role turned out to be an under-resourced experiment of sorts, and I was just another mouse furiously trying to press the pleasure button,[8] ignorant of everything I was willing to give up for one more press of the button.

I know from my own experience and from the many leaders we talked to while writing this book that there's a weight lifted from your soul in realizing that the job did not define you. After hanging up from the call about losing my role, I walked downstairs feeling somewhat ghostly. I was there but I wasn't. I couldn't process what had happened. I half snapped back to reality when one of my kids ran up asking me to play with them. "Yes," I said. For months I'd been expecting that call. I was expecting my world to fall apart, yet here I was playing Uno, and this little human was smiling back at me. I remember noticing how sunny it was outside.

Chapter 2 >_I Bet You Think This Chapter Is About You

In retrospect, I should have saved myself from that job the same way I had to save myself from red wine. Our careers are fairy tales, but the part we're getting wrong is that *we* are the main characters. We have the power to—and it is our sole responsibility to—save ourselves.

There are five seconds left on the clock. Your team is down by one and within field-goal range. You, the kicker, miss the field goal. Your team loses. It's your fault, and the ESPN highlight reel agrees. Is it, though?

That was last week. You have a game on Sunday that you need to prepare for.

> _Chapter 3

Nobody Knows Shit About F*ck

One lazy summer afternoon, as I lay on my back trying to distract myself from the day's pain, I found myself sucked into an episode of *The Dan Le Batard Show*. This one was called "Lost in Translation: Why You Can't Understand the NFL." It was the kind of title that made me stop midscroll, because honestly? It felt like a metaphor for everything in life. Football jargon, life jargon, tech jargon—same thing, right? But here's the kicker: The show wasn't just making fun of the jargon. It was diving into how it's all a performance, a way to look smarter than everyone else while leaving everyone else completely lost.

Pablo Torre, a host, was riffing on how NFL analysts like Dan Orlovsky use these impossibly intricate breakdowns that sound brilliant but leave most people nodding along like they understand . . . when they definitely don't. "It feels like everyone is just pretending to know what's happening," Pablo said. And Nate Tice, the former quarterback on the show, didn't disagree. He explained that coaches created this ridiculous language ("Spider Y 2 BANANA!") not to simplify but to complicate—to show off, gatekeep, and maintain their edge.[9] It was one big show, and everyone was playing along.

By the end of the podcast, I was equally entertained and enraged. Wasn't this just like work meetings? Or parenting advice? Or, god help us, startup culture? So that evening, as my husband and I cooked dinner, I decided to test out my epiphany. "See," I said, slicing vegetables way too aggressively, "everyone's just winging it. It's all a bunch of made-up jargon designed to confuse the rest of us while they pretend they've got it together."

That's when Ty, my partner, who is as country as cotton but smarter than your favorite *Harvard Business Review* article, stops, looks at me, and in his best Matthew McConaughey drawl says, "So, nobody knows shit about fuck."

Well, yes. That's one way to put it. After all, John Wooden told us it's what you learn after you know it all that counts.[10] I wonder: If Pablo had closed the podcast with that line, would more people get the message?

Be the Most Curious Person in the Room—Not the Smartest

After this chat in our kitchen, my thoughts time traveled back to 2014 to a conversation I had with Ted, my first boss after finishing my PhD in computer science. Ted Tanner was part Silicon Valley techie and part laid-back surfer dude. He's surfed every tech wave, from APIs to blockchains to generative AI. Ted is known to saunter into the office each day around 10 a.m., barefoot, sipping his flavor-of-the-week postworkout shake, already deep in a mathematical discussion with someone.

Chapter 3 >_Nobody Knows Shit About F*ck

During our first in-person meeting, Ted hit me with, "So, what's the most important thing you learned from your PhD?"

Immediately, a comic from Dr. Matt Might popped into my mind.[11] Dr. Might is known for accidentally becoming a pioneer in precision medicine and for the positions he's held at Harvard Medical School and the executive branch of the White House.[12] But in my conversation with Ted, it was Dr. Might's witty and realistic point of view that made him memorable. His comic sketched exactly how I felt after six years in graduate school: that my contribution to my super-specific field in our vast world of knowledge means nothing.[13]

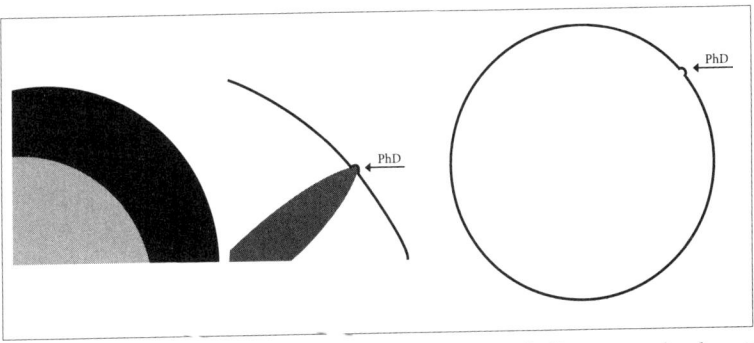

Dr. Might's illustration of what your research feels like (left), compared to how it fits into your field (center), and what that means to the world of knowledge (right). Warning: Most researchers get stuck on the left.

I remembered how I felt seeing that comic for the first time and thinking about how it took me years to make that little bump. Regardless of how significant my contribution felt to me, I'd grown to see how immeasurably insignificant my research was.

I responded to Ted's question with, "What I learned is that I know nothing. It took me years to master one problem which, in retrospect, is immeasurably insignificant."

Yes. We're all thankful that I already had the job.

But here I stood in my kitchen, ten years later and six months into healing from a work-induced mental and physical breakdown, realizing that Dr. Might's comic, a famous sports podcaster, and my husband are all saying the same thing: Our "world of knowledge" is actually quite small, and what we know about the world is infinitely insignificant.

That's the big secret, y'all. Rule numero uno. Nobody really knows what is going to happen or knows every aspect of a topic or has the singularly correct point of view. More often, folks are applying lessons they learned to a problem space that has changed radically since they learned the lessons. It is a fixed mindset that believes history repeats itself, and a growth mindset that knows history doesn't actually repeat itself, but it does rhyme.

In a universe where literally nobody knows everything, or anything with total certainty, curiosity is the only compass that really matters.

The actual smartest person in the room is the most curious person in the room. They'll stop the conversation to ask how the Spider Y 2 BANANA play works, who thought it up, and how it got its name. They use words for specificity and explanation, not gatekeeping. This is the difference between needing to show your mastery of a tiny blip of knowledge versus choosing to constantly expand it.

The Beginner's Mind: An Antidote to Imposter Syndrome's Anxiety

Not knowing everything is one thing. How you react to this information is another.

Chapter 3 >_Nobody Knows Shit About F*ck

For all of you who are overwhelmed by navigating vast landscapes of the unknown, we are with you. But here's the thing: There is a tiny tweak to your perspective that can help you skip this side quest along your professional journey instead of climbing an infinite mountain of anxiety.

Take Kat, for example. The first time we met in person, we were both middle managers at a startup. We found ourselves awkwardly shuffling around one of those tech event after-parties, sharing a tall cocktail table with another engineering manager, Shannon. Kat had just unsuccessfully tried to convince me to join the crew on the dance floor, my qualm being butt shaking with colleagues, and so we posted up over some drinks. Shannon had just dropped a bomb on Kat: "Have you ever had imposter syndrome?"

I leaned in. I felt like I was about to get the playbook for making it in my new job.

Kat's response was quick and confident: "No, I'm just 100 percent self-aware that our cofounder, Ellis, is a brilliant genius and I am not."

I was impressed. To me, that sounded exactly like someone who had a grip on reality. Meanwhile, I was already consumed by a fear loop of looking like a fraud in front of my own colleagues. I was deep in cycling through my carefully prepared list of conversation topics when Shannon jumped right in to challenge Kat's thinking. (How do people do that? Just say what they think right then and there?)

Shannon explained how Kat was comparing herself to Ellis's greatest strengths and dismissing her own unique value. Shannon pointed out that even Ellis couldn't be brilliant in every single area, reminding Kat that everyone has their own unique skill set.

She emphasized that Kat was selling herself short by magnifying Ellis's strengths and disregarding her own. At this, Kat seemed almost relieved, like someone had finally given her permission to acknowledge her talents. She nodded slowly, as if seeing the gap in her perspective for the first time.

And there it is: the twerk—I mean, tweak. You have to let go of the belief that other people have all of the answers. No one knows the universe of knowledge. Everyone has their piece to share with everyone else.

As Shannon told us that evening, "Imposter syndrome only exists because you think other people know all the stuff. They don't. Stop thinking that."

See, it is labels like "brilliant genius," "rockstar," and "smarter" that make you feel like there is something important that you don't know. Instead, walk into every situation saying in your head, "In the grand scheme of things, they don't know everything, but they do have their blip in the knowledge bubble. I wonder what they know about this?"

Take it from us. Imposter syndrome is a mountain you don't need to climb. It's a made-up place within a vast landscape occupied by the ultimate insecurity-protecting judge, your ego. (Or this could be insecurities, just one short step away from ego.)

Instead, we believe in approaching new situations tuned in to curiosity.

Chapter 3 >_Nobody Knows Shit About F*ck

Curiosity: How to Rock a New Role Without a Manual

Steamrolling people with arrogance doesn't mask inexperience, but showing curiosity can.

If there is one thing we can all agree on here, it is that remaining open and curious while walking into a new job is hard. A new gig is basically jumping right into a late-spring waterfall of Spider Y 2 BANANA riddles on repeat.

And honestly, that's all a new tech job is: continuously deciphering new puzzles and figuring out how everyone around you is building invisible stuff without crashing into one another.

From my experience, anyone walking into a new role has one of two thoughts:

1. "I know everything, and I'm finally being recognized for it! *Let's rock!*" or
2. "I know nothing about this! I wonder how long it will take for everyone else to find out and see that I'm a fraud?"

Regardless of what side of this poop coin you are on, you are wrong. I got this so wrong I ended up blowing my back out, but more on that later.

Instead, let's revisit Abby Kearns, our reigning champion over toxic ego from the last chapter. Abby is the kind of person who got hired as VP of strategy, only to blink twice and land in the CEO seat eight months later. Suddenly, she was wrangling an open source foundation and corralling a board of enterprise bigshots, all while deciding if she even knew which end was up.

So what did our newly minted CEO do when her "holy-crap-I'm-in-charge" panic set in? She hit the big red "Ask Questions" button:

- What are your best practices for [x]?
- What is the structure for [y]?
- How do you think about this?

If you are like me, deciding to speak up or ask a question in a meeting kicks off a twenty-three-minute pregame ritual. I mentally rehearse the question seventeen different ways, wrestle the fire-breathing dragon in my throat, then finally attempt to thread myself into the conversation. The whole rigmarole gives me elementary school vibes from trying to jump into double Dutch and not faceplant.

That's why I love Abby's approach: It gives me a mental anchor—"best practices"—to get out of my pregame loop and back into the room. What she's saying is that *how* you ask a question is a dead giveaway of your confidence level. Abby didn't ask, "How do I do this job?" because that's a question soaked in insecurity. Instead, she framed her questions around best practices, which assumes the knowledge already exists and she just needs to tap into it. That small shift in mindset takes you out of "I'm lost" mode and into "I'm learning" mode, which is exactly what makes people want to help you instead of wondering if you belong.

The trick isn't faking confidence—it's using curiosity to replace doubt with action. Works great when jumping ropes, too.

It's not that executives know everything. They don't. They just have the confidence to approach a job they've never done before and think, "I just need some best practices." That is a no-ego thought.

Chapter 3 >_Nobody Knows Shit About F*ck

Who cares if you don't know it? The key is assuming that you *can and will eventually* know it.

Abby didn't pretend to have all of the answers when she suddenly found herself in the CEO seat. Instead, she asked tons of questions and leaned on the people around her.

In her words, "I am, at my heart of hearts, a deeply lazy person. And so I would like to be told the most efficient way to solve a problem, glean what other people are doing, and kind of pull them together with my own personal best practices." That playbook is all she needed to quickly get up to speed and lead. Four years later, she led the company through to a successful acquisition.

Any role within Silicon Valley is a multifaceted beast. It's technical and it's business acumen. It's compassion and motivation. It's operational this and emotional that. You can make a spider diagram of your fit for a role and quickly see that the web is wonky. Why? Because nobody knows everything. You start off knowing some stuff, but not every freaking thing.

Changing your mindset changes the way you communicate. If you're insecure about what you don't know, you'll think, "I have no idea how to do this job," and you might never ask for the advice that could actually help you do the damn job. But if you replace that invasive thought with, "Nobody walks into a job like this already knowing how to do it. Even if someone had this role before, the job is completely different at a new company," then suddenly you're asking, "What are the best practices for this?" instead of spiraling in self-doubt.

Ever heard of the Peter principle? "Managers are promoted to their highest level of incompetence." The first time I heard that, my ego took a major hit. I had believed I was in tech to climb the ladder and lead. But if people are only promoted to when they become incompetent, what does that say about their intelligence?

Here's the answer: Intelligence isn't knowing everything. It's knowing how to ask the right questions. *Curiosity* is what gets you from A to B. It's how you take a lopsided spider diagram of skills and weave it into something strong enough to lead.

Knowing when and how to ask for help is the real superpower. Always be curious. That's how you own the "I know nothing" side of the coin.

Dunning-Kruger: The Other Side of Imposter Syndrome

With your curiosity hat on, imagine spending years worried you'd be "found out" for not knowing enough. Then, by changing your environment, you wake up the next day believing you know it all. This isn't far off: People switch jobs all the time.

Someone tuned to curiosity will explore what experience from imposter land transfers over into a world with confidence. Someone who's *not* tuned to curiosity might overindex on their old wins, thinking a past triumph automatically means they've mastered this brand-new arena. Welp, the latter was me, and I still cringe at the realization.

After my short executive stint (plus a PhD, a book, a few patents, and a decade working in graph technology), I started feeling more than a little confident. My special sauce of childhood influence and competitive drive whispered "FAANG" into my ear as the next step for me at this point. From there, I ~~walked~~ ~~strolled~~ ~~strutted~~ marched

Chapter 3 >_Nobody Knows Shit About F*ck

into a product leadership role at the world's largest cloud company thinking, "Hey, I know everything. I've got the combo: exec experience plus a PhD in exactly this field. Let's rock!"

And that's precisely when the Dunning–Kruger effect slapped me upside the head.

> # DUNNING-KRUGER 101
>
> When you know just enough to think you're a pro, but not enough to see the entire iceberg lurking underwater.
>
> Imposter syndrome and the Dunning–Kruger effect are basically two sides of the same warped coin:
>
> **Under imposter syndrome**, you see the enormous scope of what you don't know. You assume you're worthless, even when you're not.
>
> **Under Dunning–Kruger**, you don't know enough to realize how vast the "unknown" really is. You think you're nailing it when, well, maybe you're not.
>
> Both revolve around the same glitchy self-assessment engine we humans have: Either way, we overestimate our competence or we way underestimate it. Same coin, different side.

Well, I ran full speed ahead into FAANG trying to own the other side of that coin, completely unaware that this new arena had new rules. The first of which? Don't have this coin in your pocket, at all.

The Rift That Broke Me

I joined a massive tech cloud environment with all the swagger of someone who's totally got it. Then I met Brook, a principal engineer who had strong opinions and years of experience in the belly of this corporate beast. Our first meeting was like two Viking rams butting heads. Instead of leading with curiosity, like Abby did, I barreled in with "Well, actually, in my last role we . . ." And Brook, bless his heart, responded with an equally stubborn stance.

Because I was trapped in Dunning–Kruger land, I never once paused to think, "Maybe I could learn something from him here?" I also didn't fully get that operating in a giant FAANG company isn't the same as running a ragtag startup. Now, I see that a key difference is this: You either build something new or change something old. Not both.

Over time, the dynamic between Brook and I set off a chain of events that became a total rift in the team. The horror stories others shared about him only powered up my inner warrior: *Fix the status quo!* Meanwhile, we were building new products to catch up in the big AI race, but I was busy battling the system instead of just sitting down and asking, "Hey, Brook, what's your perspective here?"

Look, I'm owning up to my side of where I went wrong. But here's the thing: Some environments are just toxic for certain people. Those places feel like trying to swim upstream with a backpack full of rocks, every day. Being the lone voice flagging toxicity doesn't guarantee you'll spark a turnaround. You may be right, but you are alone. The confusion and disorientation can be so blinding that you miss the painfully obvious signs screaming, *"It's time to bail!"*

Chapter 3 >_Nobody Knows Shit About F*ck

Ironically, when you start asking smart questions and really digging into a topic, you might unintentionally spark someone else's imposter syndrome. In those moments, and in an effort to hide their own insecurity, people may try to make you feel like you can't trust yourself or your data. Again, it isn't about not knowing the answer. It is all about how you or the other person react. In those moments, be vulnerable. Say that you don't understand their perspective. And listen.

Next time I run into another Brook, I will be ready to respond with, "That's so interesting. It's so interesting that two people at this level have such differing opinions. It sounds like we need some more data." With Brook, leading with curiosity would have shown me if he was just an asshole, or if I really didn't have all the information. Or, as it turned out to be, both.

So don't be like the old me. Be like Abby. Learn to ask questions, learn to listen, and don't assume that just because you survived one tech battlefield, you know how every arena battle is fought. If you picked the wrong gig, you can still change course. You don't owe anybody your burned-out self.

Here's the thing: Whether you're gearing up to join the tech world or just here for the stories, you've now learned the number-one rule—nobody really knows shit about fuck. We've been there, thinking we had to prove ourselves, doubting our abilities, and mistaking others' faked confidence for knowledge. But the truth is, everyone's just winging it, and that's okay.

Get curious, get clarity, and if you need to, get out.

Find a mentor who unlocks doors and teaches you to pick your own locks.

> _Chapter 4

This Chapter Is About You

Sometimes, your passion can have such a strong hold on control that you get completely disconnected from reality. Be it your pursuit of success, recognition, or knowledge, your ego brushes off escalating warnings from loved ones that you need to slow down. Your ego ignores pleas from your body to change your situation. Sleeping problems? That's just the beginning.

It's like that story about the person who cried over and over to their god of choice to save them from drowning. After three boats pass by offering help and are denied, the person drowns. When they meet their maker they plead, "We need to have *some words*. Why didn't you help me?" Never with rolling eyes, the reply from the savior that let them drown is: "But I sent you three boats."

Watching someone you love go through this, refusing life raft after life raft, is just as harsh.

Whether it's you or someone you know, the only way to break the cycle is by recognizing the warning signs before drowning. And when we've all learned to ignore warning signs, climbing into the boat for god's sake feels impossible. My ego won't let me forget the year I lost, sidelined from my career just as my niche knowledge became mission critical. The years prior, I had ignored every signal—headaches, back pain, food allergies, panic attacks—until my body finally capsized for me. The trick isn't waiting until you're gulping water to grab a lifeline. It's admitting when you can't outswim the tide and actually getting in the damn boat before it's too late.

Here's your boat: Is there a warning light on in your car's dash, or have you been delaying a software update for a while? What about that hard conversation that you've been pushing off and are kinda pissed we just reminded you about?

Then this chapter is about you.

When you finally tune in to the warning lights of your health and relationships, you'll start noticing the ones flashing in your professional journey, too. Don't overlook them. The next time a shiny new role tempts you, remember: If it starts badly, it ends badly.

Before taking on a new role, you need to be able to say "Yes!" to the following three questions. Otherwise, you're starting badly.

1. Do I want this job?
2. Will I have the authority required to be successful?
3. Will I have the resources needed to be successful?

Now let's see how quickly things can go sideways if you skip any of these steps.

Chapter 4 >_This Chapter Is About You

Do I Want This Job?

"Wanting" a job does not mean wanting the title or the benefits or the stock options that come along with a new job. You actually have to want the job. You have to want the hour-by-hour pattern to life that comes with it.

You also need to believe in the company, the team, the product, and the market. Start by backing up from the product and thinking about how your passion and values align with the change this company seeks in the world.

Write it out:

"If this startup democratized access of _____ to _____, that would change the world, and I want to be part of that."

Next, you need to confirm that others believe in it as much as you do. For a very early startup, this might be confirming that the need for change really exists. If I was considering an early role at Peloton, I'd have asked my cycling friends if getting into a spin class at the gym was hard. I'd have asked if they're going to the gym more for the social aspect or to work with a highly motivating instructor.

If the company already had some traction, I'd want to understand how the product (or project, if it's open source) is gaining momentum. If the company has funding, I'd want to know from whom—I'm about to do some forensic analysis on the rest of their portfolio because I need to know if they have a track record of picking winners.

Will I Have the Authority Required to Be Successful?

If you need to drive change but lack the authority to do it, pass. If you're taking on a cross-functional role but lack influence on the cross-functional teams, pass.

The scope of the job to be done must match the scope of authority you have within the organization. Watch out for false promises on this one. "Don't worry, I'll clear the path when the time comes" is not the same as granting you authority.

When I took my first C-level position, I noticed that I was not actually on the leadership section of the website or in any leadership meetings. You might call this a red flag. I have, at times, missed red flags in my career. I did know that I would need a certain level of authority to be successful in my new role and brought this up to the COO, who told me, "We don't handle authority with titles and levels. The CEO will say your name more in all company meetings and ask your opinion in meetings. You don't need a title or to be at a certain level to have authority, you just have it."

[Narrator]: She did not have it.

If you do not have the authority to make changes, pass on the role.

Will I Have the Resources Needed to Be Successful?

If you're terrible at project management and you're taking on a big project, hiring a project manager is nonnegotiable. Stepping into a bigger position or something you've never done before? Insist on a coach. Plus, you sound more like a leader when you draw a hard

Chapter 4 >_This Chapter Is About You

line—demanding the resources, knowledge, and support needed to level up instead of just winging it.

Anything that is a central focus of the team must have a team resource. If automation is critical to your workflow, having someone on the team who can build and maintain scripts is essential, rather than constantly requesting updates from IT. The resources do not have to report to you, so don't feel like you're hoarding resources. But you *will* need dedicated resources for anything that is central to the team's success.

Can't get what you need for the mission? Pass.

Remember, if it starts badly it ends badly. Sometimes though, it starts well and still ends badly. Often, the answer to one of our three questions changes, and you do what we all do: power through or find another way. But over time, the strain builds, and what felt like a temporary workaround becomes a way of life.

This is exactly when burnout starts to leak in.

Burnout

Burnout can start slowly, like a bit of water slowly trickling into one of your godsent boats. Wearing rain boots in the boat will keep your feet dry, but eventually your boat/career is gonna sink.

Burnout, I know thee well.

I wasn't burned out when my niche technology failed. Failed projects and endeavors happen. My burnout started when my passion died. My passion died when I fired my friends and exchanged a job I loved for a title. My ego passed up every warning sign on the path to burnout, until my body forced me to stop.

My first boat sank eighteen months into my first C-level title.

For my third go in the startup world, I joined a pre-IPO startup to build out a new product division. The existing product was so complex that they needed someone with a PhD in graph theory (me) to lead the next wave. Blind to the red flags embroidered with red flags, I hired a team of specialists to evangelize the product and to ensure that customers were successful using it. They weren't.

Folks associate tech innovation as being on the "frontier," and we often forget what that means. It means that nobody has done this, whatever this is, or been here before. Leaders make the best decisions they can with the information they have, and often they're wrong. Startups are rarely alone on the frontier, and thus the industry gets to see the outcome of varying strategies employed by similar companies.

Water was already leaking into my burnout boat when we brought on a turnaround CEO.

Two months into the turnaround, we had not just a new CEO but new product, engineering, sales, marketing, and finance leaders. Their pedigree was solid and their loyalty was cultish, but that's good, right? No? Anyway, I got a call from the CEO who gave me five minutes to choose between two options: (1) dismantle the product and team I'd built and start a new business unit with that C-level title or (2) leave.

We all knew I was going to take that C-level title. He knew it, I knew it, you know it. *That's why he offered it.* I just needed to admit the product had failed, lay off my close friends, and update my LinkedIn profile to have a more executive-ish profile picture. I also had a book coming out that was based on the product I needed to dismantle. My ego was reminding me that I should change my cover

Chapter 4 >_This Chapter Is About You

bio before it goes to print. My heart was starting to break, and my brain was dancing in rain boots.

One week later, I was sobbing on my couch, dealing with the deep pain in my chest and stomach from having to fire some of my friends. These were people I had recruited to the team a year before who now were "too specialist" to remain within the changed mission. The next time I saw one of them years later, I cried as we hugged.

Together we could walk the spiral of decisions that led to my burnout. I'd cry writing it. You'd cry reading it. Then you'd go do the same damn thing. It happens so slowly that your body and mind adjust to their new normal. It's not even that you don't notice the water in the boat rising; it's that you justify it in all the wrong ways.

I thought that I was in a "new job phase" where things were just going to be harder for a while. I thought the ever-increasing workload was normal. I did understand that some of my health issues, like not pooping for a week or two, were stress related, but I blamed that stress on my lack of mental toughness and doubled down. I worked harder. I made space for work by stopping everything else I used to enjoy. "It's not going to be like this forever" rolled off the tongue so easily.

I've learned since that there are twelve stages of burnout, and the early phases are just about lighting the fire. The twelve stages were originally penned by psychologist Herbert Freudenberger,[14] whose research was informed in part by his own experience with stress and exhaustion. That must be nice . . . getting paid just to write about your own stress and exhaustion 😉.

Read the following eight of the twelve steps of burnout. Then reread them.

1. **The Compulsion to Prove Oneself:** The first stage is characterized by excessive ambition and a desire to prove one's worth.
   I stepped through this gate when I said yes to that shiny title.

2. **Working Harder:** In this stage, individuals start to push themselves beyond their limits, taking on more work and responsibilities.
   Need me to dismantle a department I know nothing about and rebuild it? Yes.

3. **Neglecting Their Needs:** Self-care begins to diminish as work takes priority over everything else. This includes skipping meals, not getting enough sleep, and reducing social interactions.
   Working from home is the efficiency gained from not taking showers.

4. **Displacement of Conflicts:** Problems and issues are dismissed or ignored, often with the rationale that there is no time to deal with them.
   Seventy-five percent of the people I worked and laughed with six months ago are now gone.

5. **Revision of Values:** Values begin to shift, with work becoming the primary focus. Personal relationships and hobbies are seen as less important.
   Now is when I finally can hear those escalating questions and concerns from my husband.

Chapter 4 >_This Chapter Is About You

6. **Denial of Emerging Problems:** Individuals in this stage may become intolerant and blame others for their increasing stress and dissatisfaction.
   This is all the new CEO's fault.

7. **Withdrawal:** Social withdrawal becomes more pronounced. Individuals may isolate themselves, avoiding interactions with others.
   Those friends I used to see three times a week? I don't have time for their games anymore.

8. **Odd Behavioral Changes:** Significant changes in behavior, such as becoming cynical, irritable, or aggressive, begin to manifest.
   I hiked 874 miles on the Appalachian Trail. I once woke up on a Monday morning in my tent only to be showered and on a call by ten, after hiking six miles.

While on the trail, I had *so much clarity* that I decided to quit my job and join a FAANG company so I could work with some of the other overclocked misfits I'd recently laid off.

Little did I know that my second lifeboat was captained by pirates.

I took the FAANG job knowing we had one year to launch a major product at their biggest event of the year. Six months into this journey, I was waking up with panic attacks, and one night they were so bad that I thought my heart was exploding. On the way to the ER, my husband and I said our goodbyes. But I didn't die, and instead, the doctors prescribed medication and therapy as the solution to my nighttime scaries. So, back to the grind I skipped.

Shortly after, I started becoming allergic to different foods. First, it was gluten, then dairy, then peanuts, sesame, chocolate, and rice. I continued to ignore the signals of stress-induced issues and even stopped being able to poop on my own, again. Guess what? There's meds for that too, so I thought: *This must be normal.* Great news, folks: Allergies can be reversed, and now I can eat chocolate and chicken over rice to my heart's content.

All the while, I'm thinking: *Just get past this milestone and you will be okay. You love what you are doing.* Believing that crossing a finish line was the answer, I continued to ignore my own body's cries for help. I was so disassociated from reality that I couldn't see what was burnout and what was an insanely toxic environment. I had no clear perspective on anything.

Which brings us to steps nine through twelve:

9. **Depersonalization:** A detachment from oneself and one's work begins to occur. This often leads to viewing others as objects or numbers rather than people.

   I convinced myself feeling numb and directionless was a good thing. I could just flow with the thundering herd over the cliff.

10. **Inner Emptiness:** A feeling of emptiness and meaninglessness sets in, leading individuals to engage in activities to fill the void, such as overeating, drinking, or other unproductive habits.

    My exercise and workout routine was exchanged for stuffing my face with whatever we had, late into the night. I didn't go outside for weeks.

Chapter 4 >_This Chapter Is About You

11. **Depression:** Deep depression, feelings of hopelessness, and a loss of purpose emerge.

    My husband and I cried together on our tenth anniversary hiking trip, thinking it was us, not my job. Good news, folks: Love always wins.

12. **Burnout Syndrome:** The final stage is burnout itself, characterized by physical and mental collapse, which may require medical or psychological intervention.

    I was on a call with Brook's boss at a standing desk. Brook didn't follow through with a commitment while I was on that trip with my husband. This was the straw that broke the camel's back. My right leg went numb, and my recently herniated discs weren't healed enough to keep me standing.

I went down and my second boat sank. I asked for medical leave and jumped into my third. This boat has been as dry as Oklahoma ever since.

I don't fully know when (or if) burnout #1 ended. I do know that burnout #2 started when my permissive workstyle stepped into FAANG and eventually ended during six months of rest and doctor-encouraged disability leave. My life was brought to a halt and fully deconstructed as I recovered from two annular fissures (when your discs slice open and leak), and what my medical team documented as "work-induced PTSD." It was hell, but I refused to let go of my positive attitude.

Spending too much time in "fight or flight" (or freeze or fawn) wrecks your body. The constant flood of stress hormones leads to debilitating anxiety, dissociation, and real physical damage. I learned this the hard way, flat on my back for three months, listening to books because I couldn't even hold one up without my hands and forearms going numb.

Dr. Bessel van der Kolk's *The Body Keeps the Score* should be your first stop if you suspect burnout. It doesn't just tell you stress is bad. It explains how your nervous system hijacks you, why your body keeps receipts on every unresolved trauma, and why "just pushing through" is the worst possible strategy.[15] I would know. I couldn't make a cup of coffee without feeling like I'd collapse, all thanks to a burnout injury I ignored for too long.

Every day since has been better. In every interview we did for this book that included a story of burnout or failure, we asked the person if their life got better or worse in the weeks following the event. Better. Every time. Better. In the months or years that followed? Better. Every time. Better.

> **INSIDER TIP:** The first time you tell yourself, "Well, I always did hear that it's lonely at the top . . ." You're on step 7. It's time for action.

Chapter 4 >_This Chapter Is About You

How to Build Your Own Boat

Some people naturally take action to fix their situation earlier on. John is one of those folks.

John was an early Microsoft employee, and although his team was moved and absorbed into new orgs on a regular basis, he had the rare opportunity to have the same manager at a startup for three years. *Three years*—and this was when Microsoft was doubling its revenue every year. John was elated to have the opportunity to set the vision for a major product at Microsoft. He rocked it. Then came the real work of writing product specs and requirement documents. John hated it. He didn't just hate it; he couldn't physically make himself do it.

John did not take this as an opportunity to think that he was broken and needed to buckle down. He just thought, *Fuck, I hate this*, and threw a challenge flag. He went to his boss and said, "We have to find a new role for me." Together, they went to their program lead, who after listening said, "John, you should work for me." He saw what John was just starting to see: John was at his best when working on the strategy and leading teams to execute on the vision. Taking a promotion was better than quitting or getting fired, so John was pretty pumped. Shortly after taking the role, he noticed a portion of the org that was in meltdown mode. He thought, *This would be a fun one to fix*. He requested the opportunity to turn that team around and his new manager obliged with a quip: "You're unlikely to be able to make it worse." This was the first of many teams and projects that John got back on track.

John had the forethought to see that if he'd kept trying to write those product specs himself, the higher-level vision he'd created wasn't going to come to fruition. He also just hated it and didn't tell himself what many of us do: "Work harder" or "Figure it out" or "You're about to miss your shot."

All too often, I see other people dismiss the warning signs of burnout, thinking, "I'll just push through." And I'm just as guilty here. My own bow out from the FAANG grind came after pushing too far, ignoring too many red flags, and letting my "passion" blind me to the toll that my work had taken on my health.

Chapter 4 >_This Chapter Is About You

GUARDRAILS FOR YOUR SOUL

We use humor in telling our stories (we hope you noticed 🙂), but there's very little that's funny about pushing your body and relationships to failure.

These are the guardrails I constructed during six months of medical leave:

1. **Listen to your body:** Honor your limitations. You're not an electric car: You can't just plug in overnight and be good to go. When your body starts throwing tantrums, take the hint.

2. **Be honest:** Focus on what truly matters to you. Let's be real: No one's going to give you a medal for burning out. Instead of chasing every shiny title or project, ask yourself what you really want to do with your time. Honesty with yourself is the first step to sanity.

3. **It's just a job:** Don't let it consume you. You should find something you enjoy doing with your time most days of the week. You can only work so long in a role that isn't fit for your soul's song. When work becomes more pain than passion, remember to bow out before you burn out.

Level 1

The theme here is clear: You can only work for so long in a role that isn't a fit for you. As Dr. van der Kolk's life's work demonstrates, our ability to connect with our inner reality is crucial. How can we make decisions or take action if we can't understand what our bodies are telling us? It's not just about chasing titles or working harder; it's about setting up your life to focus deeply on what truly matters.

The valuable skill isn't in pushing yourself to the limit; it's in knowing when to stop, to breathe, and to honor your own limitations.

So, as you navigate your career, remember to listen to your body, to be honest with what you want to do with your time, and to find joy in those activities that let you let go.

In the end, happiness will come from how well you align your work with the rhythm of your soul, not how long you endured a shitty situation.

Chapter 4 >_This Chapter Is About You

Level 1

Attention streaming giants and indie producers alike: I accidentally lived the greatest miniseries you haven't made yet. DM me for rights; my poor choices are ready for prime time.

>_Level 2:

Squad Goals

> 5: Overclocked Misfits
> 6: Show Me What I'm Working With
> 7: Unsung Heroes

Being aware of your ego and confronting it ignites a butterfly effect of change. If your ego still wins, you will celebrate°a new life you've built for yourself. If you tame it, you open doors to weaving together groups of wonderful souls that create new, beautiful things.

After you've begun to get your own shit together, it's time to face the next level: understanding your team. Because let's be honest, no one succeeds in this industry on their own. And we're all just as quirky as that thing you did yesterday that you hope no one ever hears about.

We're all mismatched puzzle pieces here: Drivers who can't stand a messy doc, Analyticals who crave perfect data, Expressives who think we're chasing artificial general intelligence (AGI) next week, and Amiables bringing the calm (and possibly puppies) to Zoom calls. Throw in the unsung heroes who stock the two-ply and keep your boss from looking like a total wreck, and you've got one hell of a team. Sure, the highway's on fire, half the code's held together by duct tape, and the occasional meltdown might be from you, but that's exactly why these squad goals matter. Because this crew of overclocked misfits? They're the only ones crazy enough—and brilliant enough—to make tech's wildest dreams come true.

> _Chapter 5

Overclocked Misfits

Who's creating the chips and apps on your phone anyway? What extra-sticky being lurks in a dark room late into the night, making the entire internet run?

You know the big names. They make headlines every day trying to etch their name in history. They buy islands or companies or entire social networks in an effort to be heard.

But who is actually writing the code?

Is it the young, ambitious, MBA type who is stretching their quads as they climb the corporate ladder? Probably not.

If the engineers are anything like us, the person behind the screen is a highly anxious, hunched-over meat bag with an overclocked ADHD brain. Introverts motivated by money, the satisfaction of getting a puzzle right, and a masochistic work ethic.

Creating technology done well is a beautiful thing to experience. It is a high-speed, collaborative typing contest that is most similar to repaving a highway full of burning vehicles during rush hour. You hear tales of unicorns or the "10x engineer" who pushes out more lines of code than seem humanly possible, thinks through all edge cases that could have bled the company millions, and never forgets to shut down their infrastructure. These positions require perseverance and humility. A good engineer never builds their own test suite, works with management to solve problems, and never commits uncommented code.

What most people don't get about software engineering is that it's not at all about the most clever solution, the most condensed code block, or designing fancy easter eggs—that, presumably, was all arranged before she sat down to code. Software engineering, the real business of creating, is about helping humans get something done all while juggling twenty grenades and dodging burning cars in rush hour.

Now, when we are building the next piece of *magic* to be battle-tested by the masses on our burning highway, the last thing that a manager wants is a rogue engineer. The one that sorta solves the problem but went on a side quest to hide hidden lore within an API call. You can't bring those types of engineers onto a call with a customer. Who knows if we'll be discussing the build or their philosophical discoveries from their latest desert shroom joyride. (Ain't happening again.)

Managers are looking for owners. Because someone with ownership material will never ask for a raise; they will just raise the bar themselves! *Ha!*

Managers like owners because the most elite engineers are those you can count on when shit hits the fan. Which can happen at any moment. Working in tech means that we are expected to solve problems 24/7/365. If Netflix or Snapchat or *your bank* isn't available when you need it, you start tapping harder. You probably go straight to support chat, demanding access or answers for whatever is causing your app to not work *right now*.

Chapter 5 >_Overclocked Misfits

Customer service doesn't stop for holidays or big celebrations—in fact, it gets busier. Customer service also doesn't stop for sleep, bus rides, waiting in line, or any inconvenient moment throughout your day where boredom is immediately satisfied with a brain hit from the slot machine in your pocket. Instead, a page gets sent and routed to one of us, and our version of emergency room triage begins.

When working in the triage room of the day's latest *big customer issue*, the absolute worst person to be on the call is the Academic. Academics possess egos needing to be constantly stroked, and right now the mob is screaming at us to restore service.

We need a patch for heaven's sake, not your dissertation. I don't care if the feature you want to build in five years won't work right if we take this path; the service is down and we are losing a lot of money. I don't want an elegant solution that you *hypothesize* will also fix ten other bugs from the past five years. We need to know what about the service is down, how to reproduce it, and the fastest path to getting it resolved. Do we just roll back the latest deployment? Probably.

And look, trust me, I was the Academic. I spent years obsessing over the perfect theoretical fix, arguing over optimizations that wouldn't matter for a decade, and clinging to intellectual purity like it was a life raft. But here? The only thing that matters is putting the flames out before the business loses another million dollars.

After all this, let's have a proper RCA (root cause analysis), where we can go into options A, B, C, D, and Z. But right now? Go on mute.

Give me the engineer who grew up learning on our code base and I know they will make sense of our ball of duct tape. And don't be fooled: Everyone's production code is one strip away from becoming unglued, and only Brian knows which piece that is.

The engineer who joined partway through the project? The one that boasts that the latest obscure trend in tech is exactly what we need to refactor our code base? They can sit this one out.

Ultimately, we do attempt to balance production issues with R&D and deciphering if there's any science in the latest tech wave. Those types of problems are perfect for a deep thinker. But we don't have a dissertation's amount of time to figure it out when production is on the fritz. Come back to me in a week with a six-page position on the trend of the week and be clear on why we won't be doing it (if you are at a tech conglomerate). But if you are at a startup, make that six-pager about why the latest thing is our next company pivot.

There are plenty of exceptions to when Academics shine in production engineering, of course. And I know I'm not one of them. I worked with a PhD scholar from Georgia whose communication and speed to deliver were top-notch. With six months to go ahead of a product launch, he pitched and got approval to add a major feature to our product that made it relevant to the AI wave that we were already a year late to catch. Once it was done, he joined me in speaking with customers about it and made me feel like a colleague instead of another middle manager to dismiss. Give me more Academics like that: That's how to walk from a university into the biggest of machines and be the epitome of professional engineering.

One of the most coveted roles to find in tech is being a "strategic technical advisor." There's a dozen different titles for this, so that's not what to look for. This person reports directly to some C-level and has a title that fits just enough for HR to not ask questions. They've

Chapter 5 >_Overclocked Misfits

probably been around a long time because they are one of those unicorns or 10x engineers who are now burned out. Their job is to be an interpreter of whatever latest buzzword is trending on Hacker News that week. They are the sous-chef, with the end results being just as tasty as the chef's (or not).

The advisor has two modes: translating the latest tech trend into money or chillin' and smoking bud. And the C-level is just fine with this arrangement. Leadership doesn't want to sound like an asshat during their latest customer meeting or podcast recording. Knowledge is an executive's currency, and careers start to crumble when they look like they don't have any. As a result, leadership puts up with the advisor's late-morning starts and backpacking escapades as long as they keep seeing through more bullshit than everyone else. So, light up and catch those rays, we'll call you in when we need ya.

Being a woman in engineering is still a game of dodging misogynistic punches, but with more awareness of just how freaking many there are. In the decade since working with Sheryl Sandberg on the book *Lean In*, I can see the game more clearly and have taken self-defense classes. The key is to find your allies and recognize that not everyone is one. You can usually tell if someone is an ally or not. My only issue has been trusting my gut on the matter, where my experience says trust always leads to the better outcome.

An ally in your tech circle looks like someone who interrupts you in a meeting out of excitement but then immediately catches themselves, apologizes, and yields the floor.

Someone who isn't your ally probably held a meeting on the topic and didn't invite you. That's obvious when all of a sudden three or four people speak up in support of the same topic, going in a new direction that you've never heard of before.

Still not sure who is who? Unzip your fly and walk around the office. Allies give you a heads-up—everyone else just Slacks about it.

As an engineer, the tickets you close are your badge of honor. On the days when you actually get more than one uninterrupted hour to actually work on them, you set up your multiple monitors like an artist prepares their brushes. Code in the center, your fav gpt on the right, and your playlist set. You can feel the lock-in build in your mind. You've also ignored all of those software updates for one more day because today, you are going to get shit done.

Early in the game you open your priority queue and go for the quick hits. The rush of dopamine from picking up and closing out a few low-value tickets makes you feel accomplished. They build up some confidence.

Never underestimate the power of "closed, successful" tickets and winning a few more story points for your team.

The experienced pros do it differently. They've learned that quick hits from doc touches don't move the needle. Instead, pros apply real mental strength to select their work from the other side of the priority queue. The best engineers spend their longest time blocks and freshest mental energy working through the biggest issues. There's a trap here though. The pros still deliver on time instead of hiding out for weeks on end with no update. They do that by actually talking to product managers and providing time estimates they know they can hit. They have a realistic outlook on how many of those deep-thinking windows they really have in a week, and they stick to them.

Chapter 5 >_Overclocked Misfits

The secret to building fast is completing features on schedule. And leadership wanted to know eight weeks ago that the piece you were building was going to take eight weeks. Don't stress about asking for eight weeks; they'd rather give an accurate estimate back up the chain. Just adjust the schedule ahead of time and line up everything else that's moving according to it. Well, except for big marketing launches—but let's save that for another time.

No matter your role in shipping code, there seems to be a shared pride in building the next frontier alongside your fellow misfits.

The term "misfits" was first introduced to me by newly acquired coworkers after my stint in academia. Over lunch on my first day, the crew boasted they were an "island of misfit toys," as was evident in our office's living room.

The first "real job" I had after academia operated out of a musty former dentist's office, complete with flickering fluorescent lights and a faint antiseptic smell that never quite faded. Our CTO's executive assistant set up shop in the glass-walled check-in area and brought the only warmth into the space. The former waiting room, now our so-called headquarters, was a chaotic mix of startup grit and dorm room décor. Oversized, fart-tinted beanbag chairs sagged in the corners, a pirate-themed skeleton draped in a feathered boa stood guard over our makeshift lounge, and an ever-growing collection of booze lined the shelves like a shrine to our long nights.

We only grew out of that office when the CEO raised Series B funding. We were so enthralled to be moving out of the musty living room that the whole company gathered next to the skeleton to hit refresh on a browser and watch $20 million hit our bank account. Shortly thereafter, part of our funding went toward a red-carpet party on an aircraft carrier in the Charleston harbor. Because nothing screams "responsible use of our runway" quite like popping champagne on a literal warship, right?

However we got there, I was enthralled to have an exit from working on a card table next to the bathroom where I could smell everyone's business post-morning coffee.

To proudly claim the label of "an island of misfit toys" really meant the team previously felt like outsiders and found camaraderie because of it. We weren't the ones who fit neatly into corporate molds or followed the expected career paths. Maybe it was the engineer who rebuilt original Macintoshes in their spare time, painstakingly restoring each circuit and keyboard as if preserving a lost artifact. Or the colleague who played first chair in a local symphony orchestra, balancing late-night code commits with early-morning rehearsals, seamlessly switching between debugging software and perfecting a concerto.

These were people whose passions extended beyond the usual, who brought the same level of dedication to their work as they did to their hobbies. Their interests might have been unique and nonmainstream, but that never meant they lacked focus or drive. If anything, it made them even more relentless in pursuit of solving problems, refining ideas, and building something better.

Your job in this world of overclocked misfits is to figure out who your allies are, to stay away from the toxic culture sucker, and to not become one of them yourself. Building confidence by getting quick wins from closing tickets and being listed as an author on your friend's big paper can easily fuel weak egos.

Chapter 5 >_Overclocked Misfits

A team member who instead focuses on collaboration and understanding the bigger picture brings a different kind of energy than someone who sees every discussion as a chance to prove themselves. I'd rather work with a computer science intern eager to learn than someone who dismisses questions and steamrolls the conversation with their own vision. Skills can be taught. Processes can be learned.

But I can't teach character.

We build the engine, fine-tune the gears, and drive full throttle through chaos, all while tech's golden boys take the podium. But, make no mistake, this is who wins the race.

>_Chapter 6
Show Me What I'm Working With

I learned the hard way that most things aren't about me after a tough conversation with a leader I deeply respect. He said, "The CEO is not responsible for your mental health. He is responsible for the success of the company, and if you want to destroy your mental health in support of his mission, he will let you."

If life were a cartoon, I'd have just been hit in the face by a large fish. I needed to binge-watch TED Talks on boundaries and I needed to show up at work differently. I needed a new framing. The CEO and I were not friends, but we were brothers-in-arms. I existed in service of a mission—his mission.

The idea of servitude is a difficult concept for the prideful; however, if you can accept that things are not about you, then the only logical conclusion is that they are about the product, the company, and empowering all of the overclocked misfits, ourselves included, to punch above their weight.

I needed to know how to best serve this person that I was following into battle every day.

In this pursuit, I found the Social Styles Matrix.

The Social Styles Matrix

There are lots of frameworks out there. I bet you even know your Myers–Briggs letters by heart. The problem with most of the popular frameworks is that they're too complicated. We get our ego stroked by their fancy quiz that offers some new insight into who we are, but I often wonder if each framework's complexity is a sales tool. Management training is sold by the day, and we need to get our money's worth.

The Social Styles Matrix is simple. It's something you can read once and use for the rest of your career. It requires you to make just two observations about a leader:

1. **Ask Versus Tell:** When a decision is needed, do they start with gathering information or sharing their opinion or knowledge?
2. **Open Versus Closed:** In conversations, is it easy for them to share their feelings, or are they more reserved, even hidden?

Chart your analysis using the following image to know if you're working with an executive who's a Driver, an Analytical, an Expressive, or an Amiable. Then, you just need to know how to work with them.

Chapter 6 >_Show Me What I'm Working With

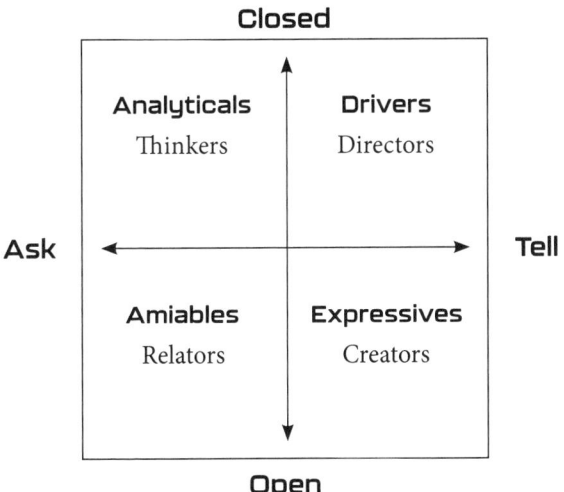

Most personality frameworks are ego mirrors. This one's a tactical map. Two questions, zero fluff, and a fast path to figuring out how not to piss off your boss.

Drivers

We're starting with Drivers because they'll absolutely lose their shit if we don't. They also wouldn't waste their currency, *time*, to hear about the other executive styles.

Drivers are forceful and determined. They are action oriented and want outcomes. Outcomes, not actions. That's an important distinction. Actions are things you do on the way to outcomes.

You exist to provide options and a recommendation, but they will make the decision. "I've thought about this a lot" is unwanted foreplay. In this game of fetch, your job is to determine where they want the ball thrown and then to throw the fucking ball. The approach is, "Your two options are x and y, and I recommend x for this very concise reason." Answer their follow-up questions with facts. This next part is tricky: You need to give them time to think through the options. If they answer immediately, it looks too much like you made the decision. Give them time and then go with their decision.

I work closely with a Driver. Believe me when I say the things they don't like are more important to internalize than the things they do. Your work must not look disorganized or messy. Beautiful data in an ugly spreadsheet will be viewed as garbage data. If they call, you answer. If you strongly disagree with a decision, align the facts you have to the outcomes they seek.

Drivers do not care how the last guy did it. Again, provide facts and align those facts to the outcomes your exec is seeking. If you say something will be done by EoD Tuesday, submit by lunch. You will miss a deadline at some point, and when you do, it's helpful to have a track record of beating deadlines. If something is taking longer than it should, extend the deadline before the deadline arrives. A string of misses will land you on the sidelines. If you're going on vacation, *say it*. Be clear as to whether you are reachable or not. Drivers will only respect the boundaries you're willing to state and stick to, and they do not like surprises.

You can set boundaries with Drivers, but you need to understand that doing so has consequences. You do not have to answer your phone on the weekend, but a judgment is being made. Instead, you need to find a balance. Let a Driver know *when* you'll be available. You don't have to answer, but send a follow-up text explaining that you're in the middle of something and will call back within X time (be precise-ish). You're still taking the call on a Sunday but not until the family lunch is done.

Chapter 6 >_Show Me What I'm Working With

Documents need a level of organization that might feel unfamiliar to you. Ask one of the operational savants for an example of a document that your exec liked. The ops team will save you, and it doesn't matter if they think you're an idiot. Look at how the doc is organized. Long form? Mostly bullets? High-level TL;DR at the top? What's the font? What's the line spacing? Are sections delineated by bolded and underlined text, or are we working with levels of headings on this bad boy? It is *your* job to fit *their* style.

I'm terrible at all of this, but on a good week I can play the game, and so can you. If you're like me, you will never be the Driver's favorite. *Stop trying.* They could care less about this, and if the favorite feels threatened, they will crush you. The exec sees your value, and that's why you still have a job. You have to do the minimum on these things, and that's probably five times your previous level of achievement. Stop there. Your exec needs you to produce the things he keeps you around for. There will be mishaps.

I missed a call from my exec on a Tuesday. I noticed it the next day. I slacked him with an apology, but he iced me for another couple weeks. Instead, his operational savants brought me his balls and very detailed instructions for throwing them. There you have it. Answer the fucking call or you'll have a team of folks bringing you your exec's balls.

Lastly, do not justify the behavior of a Driver to the person they offended, including yourself. Drivers know who they are, and they don't want anyone apologizing on their behalf. That shame you feel after they shred someone in the meeting is a you thing, not a them thing.

Analyticals

With Analyticals, the word "precise-ish" doesn't exist. If you use it with them, you will lose any potential credibility you had when you joined the meeting. You can, however, *very* sparingly use the phrase "directionally correct." It's best to duck after doing so.

The currency of Analyticals is information. *Correct* information. *Correct* information that they *trust*. They are closed but want to gather information before making a decision.

Analyticals have three states of data-being:

1. **In love with the data:** In love with the data to the point they lose sight of the vision.

2. **In hate with the data:** Untrusting of the data or to what degree you coerced the data to prove the point you wanted to make.

3. **In acceptance of the data:** Comfortable with their understanding of the flaws in the data enough to trust the data for what they believe it should be used for.

You're about to spend a lot of time creating docs and spreadsheets that are logically coherent in hopes of landing in the love or acceptance buckets. Lucky for you, Analyticals will tell you what they don't like about your logic and your data. It's a dance. There are times when preparing to deliver your findings and insights will take as long as it did to reach them. The effort is worth it. Sure, you'll end up needing glasses from the eye strain, but if you can get comfortable making Analyticals comfortable with your findings, you're set. The Drivers, Expressives, and Amiables will eat that shit up.

Chapter 6 >_Show Me What I'm Working With

Left to my own devices, my spreadsheets are ugly. I recently needed help understanding a concept and asked our chief accounting officer for an assist. He was happy to help, but during the Zoom I was like a confused turtle trying to explain my findings because I'd made my spreadsheet functions way too complicated, all because I wanted to look as capable as possible while asking for help. Finally, he stopped me to say, "You don't have to go so slowly. I can pick up the math as you talk." I had to explain that I was going slowly so I could remember what I'd done. He asked if he could drive, clicked through some cells, then informed me he understood the math and that he was ready for my questions. I asked, he answered, and we were done. I was wasting his currency. Next time I'll send the doc ahead of time so that he knows what I was thinking before I forget.

When Analyticals are in hate with the data, you mustn't take it as a personal assault, unless the data is actually great and you did a terrible job presenting it; in that case you should take it personally. Data is imperfect, and Analyticals have trouble overlooking those imperfections. Folks like me often game the data, and thus Analyticals start to distrust it. To get past this, help the Analytical understand the relationships between the data—often, this changes the conversation from month-over-month absolutes to something like conversion rates—and focus on how those improve or decline.

Take ads for an example. If a company is selling skateboards, they're expected to run ads for skateboards. The market for skateboards has limits, and so instead you run ads for skater T-shirts. A lot more folks land on the website, but they realize the cool T-shirt is free only with the purchase of a skateboard. The metric for website visits has skyrocketed. So has ad spend. But that's not the point here.

Level 2

The Analytical now knows the website visits data is shit. This is now your mess to clean up, so instead of an absolute number ("visits"), you have to refocus on conversion rates. What percentage of people that clicked on an ad went on to buy a skateboard? That's the new metric. We'll talk more about this in Grow the Flow (Chapter 11).

Analyticals like schedules . . . until they fall in love with the data gathering process. Without the schedule, they'll stay in data collection mode indefinitely. With the schedule, they could feel rushed and, with the executive power they wield, set the schedule aside. This is why VCs love Analyticals: They'll spend two years perfecting an idea before anyone ever hears about it. This is also why Analyticals can drive a CEO crazy: They'll spend two years perfecting an idea before anyone ever hears about it.

As with Drivers, you do not want to be disorganized or messy, and you don't want to rush their decisions. With Analyticals, you also need to be careful about using unreliable sources, like the person that was gaming the visit metrics with T-shirt ads.

Every exec has an Analytical alter ego. But for some, Analytical leadership is a science. A science you've just mastered.

Lastly, never use a pie chart unless it's for your exec's birthday cake. That would be hilarious. Please write to us.

Expressives

We're covering Expressives third not because they're always late (they are) but because they *love* the anticipation of the stage time. It's your moment, Expressives.

The currency of Expressives is vision. There are lots of I's in vision, and the E in Expressives stands for ego. I'm allowed to say

Chapter 6 >_Show Me What I'm Working With

that because I'm an Expressive. I felt the need to tell you that because I am an Expressive. It all started when I was five . . . just kidding. That's an Open/Tell joke. Expressives are open with their feelings and ready to tell you their opening vision quickly.

Expressives might be late to the meeting, and honestly they might forget about the meeting completely. They most likely won't come with an agenda, but they're working toward a vision and progress is expected.

When you're working with an Expressive to shape a vision, you must remember not to let your Analytical or Driver genes show. Do this by abstracting your pushback by about ten thousand feet. If the Expressive says, "Did you hear about the latest AI breakthrough? Just think about what that means!" they're not looking to hear about how it impacts the near-term road map. They want to know how it will change the future.

You have the tools in your toolbelt to respond. You know where your market is going, and you just need to think about how the breakthrough du jour will get the industry there faster. Who does this breakthrough commoditize and who just got unblocked?

If you're stuck, think about that one thing you keep hearing about that seems far-fetched at best, and lean into that (e.g., "This could be the breakthrough that makes AGI real"). The Expressive will often jump in to continue your thought, but trust yourself to map it aloud as well. Expressives don't care if ideas crash and burn. They care that you can think big enough to have the conversation.

Think big but also be curious. Be curious about big stuff, but be just as curious about dumb stuff. Einstein was riding a bike when he wondered what it would be like to chase a beam of light.

Exploring that thought, and a sprinkling of math, led to his theory of relativity.

You do not need to match the level of creative genius of your Expressive. There will be folks on the team that naturally do, but the Expressive isn't playing favorites. Often, everyone is their favorite. I once worked with an absolutely brilliant Expressive. When he joined the company to lead strategy, his first hire was a longtime associate. This guy was an obviously brilliant Amiable with a streak of Analytical that was enviable. He sought no recognition, hated it really, and just wanted to think deeply about hard topics which were often in line with the Expressive's vision. They were a killer team. The Expressive could share an idea worth shaping, and he would offer data in support of the vision and even examples of philosophical alignment from the great thinkers throughout history.

Left alone, the Expressive has no concept of deadlines. Keep them focused but never let them realize you're doing it. An approaching deadline is more your stress than theirs. You must remember who set the deadline. It was probably the CEO, who is very used to the Expressive's process or, um . . . antics.

There are times where you may not know where you sit with the Expressive. They can be hard to read. They're comfortable sharing a range of emotions, and if you've mostly been surrounded by Analyticals and Drivers, you might think the Expressive is losing it. You think this because it's only in a spiral that the Analyticals and Drivers show such ranges of emotions. In time you'll find this refreshing. What you originally saw as difficult to read is actually exactly what you see.

```
Chapter 6 >_Show Me What I'm Working With
```

Do not make the mistake of thinking that your Expressive doesn't have drive or a need for data. This person is often setting the three-to-five-year strategy for an organization; they know their shit, but they understand the value of the creative process. Add the book *Rest: Why You Get More Done When You Work Less* by Alex Soojung-Kim Pang to your reading list. Creating a vision and strategy requires an exhausting level of deep thinking. Seek to understand the process and carry some of those lessons with you.

This next piece of advice is important. Don't overrotate on the natural mentor-esque nature of the Expressives. Their desire to help and fix can become an aspect of the relationship that you overvalue. Your desire for their mentorship can lead to them seeing you as more junior than they once did. It's a tough spot to get out of, so just avoid it. If you need help, get a coach or a therapist. It's usually free with health insurance, and I jokingly call that the "You broke it, you fix it" policy.

Amiables

Amiables are the diplomats. Trust is their currency, and they spend it ending wars and influencing big change.

The Amiable is often the smartest person in the room, but they don't need you to know that. They see you getting stressed out and they find that curious. They're skillful at picking battles and masterful at making sure you don't realize they've chosen the battle at all. Often, you'll walk away feeling like the victor while doing exactly what the Amiable wanted.

It's annoying as hell, but you'll also watch your Amiable exec have a perfectly fine relationship with the one employee that drives you bonkers. Folks often underestimate the Amiable. In tech we celebrate the eccentricities of executives and write books about these quirks. You know these: Execs get up at 4 a.m. to write, they always make their bed, they require no more than five hours of sleep, they have a four-hour workday, they work one hundred hours a week, they ski but only when dropped from a helicopter. Thus, we judge the level-headed Amiable that fosters puppies and rides bicycles as less-than.

They're rarely the startup CEO but often rise to that rank at larger companies. They are, hands down, the best exec you'll ever work with, and in times of stress you should do your damndest to channel their energy.

When working with an Amiable, you want to present findings in a balanced way. Ask the Amiable "why" questions and make it known when your opinion has been swayed during the research phase of a project.

Be organized but not overly. Amiables are often gifted at simplifying decisions and, in my experience, financial models. Where the Analytical or Expressive will overthink a model that is to inform a decision, the Amiable will show up with the fewest considerations that will reach the most logical conclusion. They naturally think, "What would have to be true for this to make sense?" and they do not expand much from that simple thought unless needed.

If you're meeting with an Amiable, show up on time and never have a hard stop right after. Level-headed reasoning can take time. They're often teaching you something as well, and cutting the session off is seen as you not being fully invested in the lesson.

```
Chapter 6 >_Show Me What I'm Working With
```

If you're having trouble with another exec, it makes sense to ask for advice from the person that can work pretty well with anyone. I worked for an Amiable, and together we worked with a Driver. The Amiable, knowing this Driver very well and sensing how much I didn't, said, "You gotta bring him a rock. He needs to see the progress. Every day or two you need to bring him a new and shiny rock." It was so simple, yet it was exactly what I needed to do.

Let's Hit Replay

We asked a Driver to do an interview for this book, and they declined because they didn't have the time. The Analytical agreed to do a book interview once we provided an outline and the framework in which they would be quoted. When we got on a call with an Expressive, we asked one question and ended up needing a second call after we hit a recording limit on our meeting software. We interviewed an Amiable and . . . I don't have the heart to make a joke here because they're less deserving of the jab and I want them to tell me about their next litter of puppies they're fostering.

> _Chapter 7

Unsung Heroes

When you're secretly crying in the bathroom because the CEO shredded you in another meeting, remember that someone stocked that quilted two-ply toilet paper you're using to dab those tears. When you find the strength to leave the bathroom and collapse into a beanbag chair to rewrite your plan, obviously grabbing a can of sparkling water and organic Reese's Cups on the way, remember that those perks don't just magically appear.

When a potential customer wants to change their sales contract twenty-four hours before quarter-end and legal saves your ass, understand that they did not have to save your ass.

When similar startups with better products and more funding are folding and y'all aren't, thank finance.

When anything at all gets done that requires two or more teams working together, thank ops. Actually, when anything gets done, thank the ops team.

Engineers, sales, and even the marketing peeps are idolized in tech. I say *even* marketing because when you start out in tech, you think marketing takes "nobody knows shit about fuck" to a whole new level. You're shocked to learn that the best data scientist in the company is optimizing ad spend and the second-best data scientist is working on a machine learning model to more quickly identify high-quality leads. The real sting comes when you realize that to the outside world, the developer relations teams, who often roll into marketing, are the real heroes of a product's success. They're the ones evangelizing adoption and making champions out of users. Hell, if you look at others' shoes more often than your own when walking down the hall, you might even become an evangelist someday.

If you want to develop as a leader, you have to step outside of your bunker or cubicle corner. Taking on the insular identity of your department may feel comfortable, but it feeds the ego more than it serves the team. Leadership isn't about mastering every function; it's about understanding how they fit together. An orchestral conductor may not be a brilliant trombonist, but she sure as hell knows how every section contributes to the final sound.

The same applies in business. You may not know much about marketing, operations, or finance, but you are all wearing the same jersey. The only way to improve, to make better decisions, and to lead effectively is to listen and engage. The ego wants to assume its own perspective is the most important and the most correct, but real leadership means setting that aside. Appreciating and understanding other functions isn't just a nice-to-do, it is critical. The best leaders aren't the ones who know it all; they are the ones who know how to learn.

Listen to learn, learn to lead.

Ops

I like to think of myself as the concertmaster (head violinist) of our orchestra. I like the orchestra analogy more than the traditional "cogs in a machine" analogy because it recognizes that humans have agency and can leave, but doesn't lose sight of one's replaceability. I also like thinking of myself as the concertmaster because letting go of ego is a journey. I am, however, okay with everyone else thinking of themselves as the concertmaster as well. See? Growth!

Another area of growth for me has been in my appreciation for the operations team. If I carried the orchestra example through this chapter, you'd expect me to assign ops to drums. You'd argue that they're keeping the operational rhythm of the company, and they do, but at a higher level.

You might also expect the CEO to be the conductor, but the CEO is the composer. It's actually the operations team conducting our orchestra.

I can tell you from experience that if the concertmaster (me), pisses off the conductor (ops), they won't be the concertmaster for long. If you're about to fall from a cliff and someone hands you a rope, it's safe to assume they're helping. If you're failing in a leadership role and the ops person is *constantly* asking you for metrics, giving you advice on leading your team, and calling more and more cross-functional meetings, your ego might think that the person approaching with a rope is actually coming to stomp on your fingers instead. Make that mistake enough and you'll be right about that stomp.

The ops team, and project management folks by extension, are responsible for keeping plans on track, escalating issues, managing cross-team communication, and aggregating company-wide observability reports and metrics. When an ops person asks you to explain a metric, it's not because they care that your numbers tanked, it's because they know that the C-level staff is going to zero in on that metric in a weekly meeting, and the ops person is giving you a chance to eliminate guesswork on the leadership's behalf. They're eliminating round trips of information gathering.

I guarantee you'll be in an offsite where someone thanks an ops person, which cascades into this moment of realization for others that just about every industry accolade they've received wouldn't have happened without operations driving things forward. They're the often thankless hero makers of the org.

All this is great, but if you find yourself unable to visualize the orchestra example because only the cog-in-a-machine example fits your situation, it's not good. If employees lose their agency, they no longer feel an appreciation for their skills. An over rotation on ops for all cross-team communication can become viewed as a barrier between leadership and the rest of the organization. This feeling of replaceability seeds a breakdown in trust. The paths that ops once cleared start looking too much like an API contract. Clear the path, don't bulldoze it.

A great ops person will clear any barrier in your path. Let them.

Admins

The operations team might know everything happening at the company, but the admins know *everything*! Unfortunately, they're also vaults. Those secrets are collected and not shared. You might have to wait ten years after a company has been dissolved to hear the stories about the CEOs walking in on Monday mornings to drop piles of "expensable" receipts from strip clubs on the admin's desks. Those stories are worth the wait.

Chapter 7 >_Unsung Heroes

If you have a challenging relationship with an exec, talk to their admin and ask how you can improve the way you work with that person. They're going to tell the exec with whom they share the vault. That's their job. But when the relationship improves, nobody cares.

"Admin" is a somewhat dismissive title. Admins are the chiefs of staff for the exec or execs they work closely with. They're the personal "fixer" for that exec. They shift stories as much as they shift schedules. The biggest mistake new execs make is underestimating the superpowers of their admins. This is the one person in the company that is allowed to know that you're a wreck. A good admin will ensure that nobody else knows. They will fill the gaps in your public profile so that you look like a whole-ass person.

I also firmly believe that the number of admins that quit working for you directly correlates to how shitty of a human you are. This might be the one time when . . . it actually *is about you.*

Finance

What does it mean to "keep the lights on"? It's nothing more than budgeting, forecasting, financial modeling, managing profits and losses, keeping the financial books, preparing financial statements, managing expenses, paying the bills, paying the people, designing the bonus plans that pay the people, making sure the SEC doesn't come after you, recognizing revenue and managing recurring payments, managing the employee stock program, maintaining the cap table, raising capital (preparing pitches, doing all of the due diligence, and negotiating term sheets), managing ongoing relationships with investors and potential investors, managing relationships with the industry analysts, making sure you don't run out of cash, paying taxes, doing the required regulatory reporting, coordinating audits, understanding your financial risk, making sure you're insured, ensuring there's no internal fraud happening, providing all of the insights to support building a growth strategy, leading pricing strategy efforts, providing cost-benefit analysis, preparing for an IPO (although everything on this list counts as preparing for an IPO), planning for mergers, acquisitions, or buyouts (yours or that of an external company), reducing costs, negotiating with vendors, improving operational efficiency, managing your debt, providing models for potential investment strategies, and making sure you always have contingency plans and paths to emergency funding when shit hits the fan.

Otherwise, they mostly play golf.

FREE ADVICE

If you need something from finance within the two weeks after quarter-end, save it. They're doing whatever "closing the books on the quarter" means, and you do not want to get in their way.

When there's a company decision you don't understand, it's often because you are considering only your perspective. Think about what would have to be true for the decision to make sense and then mentally travel through the other departments. Often, you'll find that reason within finance.

Legal

The rule with finance is that you leave them alone for two weeks after the quarter ends. The rule with legal is that you leave them alone for the two weeks before the quarter ends. They're heads down negotiating contracts with customers.

That's actually just a fraction of what legal does. Every major company decision has a downstream impact on legal. Every product needs a name, and whether or not you can get away with stepping on the toes of some random trademark is up to legal. Every product needs a license. Every service needs a terms of service. Every open source project needs a license. The merits of open source licenses are constantly debated along with who is and isn't a good steward of them. These opinions are rarely voiced by the legal team, but if you really want to understand the options, book time with someone in legal.

Not every early startup pursues patents, but as the startup matures, they usually do. Using patents to create an exclusive market creates barriers of entry to that market. This is a bad strategy. You want the biggest market possible, and using intellectual property to limit it early on is just going to make your life harder. If a hyperscaler (cloud company) can easily launch a similar product to yours, your moat was too narrow and too shallow. Start over.

If, however, you want to pursue patents to attract investors and users because you want to be seen as innovative, go for it. Legal is your friend. Likewise, patents are bucket-list items for engineers, and supporting their dreams via the patent process is a great way to attract and retain talent.

Bottom line, involve legal early on because covering your ass is cheaper than defending it.

HR

HR works for the company, not you. When you need help with reorganizing your team, leveling comp, or the communication of something hard, ask your HR business partner. They sit right next to legal and understand the business risk better than anyone. They will help you not make the same mistakes they see over and over again.

And again: HR sits right next to legal and understands business risk better than anyone. If you go to HR for a harassment issue or to get an accommodation for a medical issue, they morph into actuaries. Every action is weighed and measured according to its defensibility in a legal discussion. You are now a quantifiable business risk and will be treated as such.

It happens more than you think.

Everyone Else

We didn't call out the data team, but you know you owe them your success. You can't make a decision without them, and nobody would want you to. IT is a real unsung hero, too. Can you imagine working in IT at a tech startup? It's absolutely thankless until someone like me accidentally sends an acquisition-related email to someone with the same first name as the CFO. You will quickly learn that there's an unsung hero that can delete it, only three minutes after the mistaken receiver reads it.

Chapter 7 >_Unsung Heroes

Know your heroes before you need them, my friends, and don't get to know them because you might need them. Do it because you want to foster a good culture and because they often have the key to the swag closet.

Level 2

>_Level 3:

Learning to Lead

> 8: Unicorns Are Real

> 9: Build Products That People Want

> 10: Stick the Landing

> 11: Grow the Flow

> 12: Data Isn't People

From S-curves that reveal how scarcity morphs into abundance (and back again) to flywheels that build momentum like a merry-go-round on steroids, these chapters show you how real growth happens: one push at a time. Spot a wave, ride it, and if you miss it, pivot before you burn all your resources.

You'll learn that writing down your educated wish keeps you from building the next fiasco, that a product might be impressive but still flop if it doesn't address a real scarcity, and that naming anything is as fun as finding out you're allergic to your new fancy wool coat.

Don't worry, the overclocked misfits and unsung heroes are still in the mix, but now they're strapped to data dashboards and fueling your flywheel with measured precision instead of guesswork. Think of data as your cockpit instruments. Data is not about perfection, but about agreeing on a tool to use in conversations instead of following the loudest ego in the room. Yes, you'll face last-minute heroics, cringe-worthy product demos, and enough pivoting to make your head spin, but that's exactly why S-curves, flywheels, and well-wielded data make up your survival kit.

Your gut instincts and big ideas have gotten you this far, but there's a science to finding product market fit (PMF). A science that you must internalize or you will fail. A science that includes understanding the psychology of the buyer and the forces of the market. A science that you are about to master.

>_Chapter 8

Unicorns Are Real

If someone really wants to learn to cook, they buy *Jacques Pépin's Complete Techniques*. Jacques starts by teaching you how to pick a knife. You then learn how to sharpen a knife, and once you've mastered that, he deems you ready to learn how to hold a knife.

Should you have the dedication required to keep going, he will teach you how to chop an onion with so much elegance that you'll have dinner guests watching in awe as you chop an onion twenty years later. This is not that book. The closest thing you'll find is Y Combinator's Startup School, which is free online and goes at a pace that matches the eagerness of founders.

We're going to take a step back and tell you which knives you need and how they work. At minimum, chefs need a chef's knife and a paring knife. At minimum, you need S-curves and flywheels.

Scarcity–Abundance S-Curves

What change do you seek in the world?

The change you seek doesn't have to be to cure cancer or to get Banksy to do more print releases, although both would be appreciated. The change you seek in the world can be as simple as clearing an obstacle or challenge that is making your day-to-day work harder.

I gained this clarity when I first met Ben Firshman. Ben is a four-time founder and prolific open source developer. I asked him about his motivation behind developing a specific open source project, and the question almost confused him. He said something like, "I needed that software, so I built it, and then I open sourced it because other people needed it too."

Ben certainly was lucky to face these challenges within the early phases of massive technology waves when resources were easier to come by. Multiple successful startups were born of his simple logic. Well, buckle up, because I'm gonna show you how to make your own waves.

There's a wave-generating tool called the scarcity–abundance S-curve. All software products start in the same place: Something is scarce, which is creating a problem that you want to solve. The software you develop to address that scarcity will make something else abundant, and that abundance will create a new scarcity.

Let's step through a real example:

- **Abundance:** It's 2013, baby! We have so many different operating systems and hardware options.
- **Scarcity:** There's no easy way to deploy applications to every environment without writing specific code for these differing operating systems and infrastructure options.
- **Abundance:** Docker containers address the scarcity with lightweight, platform-agnostic packaging of applications. There are one million container downloads in year one.
- **New scarcity:** AHHHHHHHH, it's 2014 and there are one billion container downloads! How do we manage all these containerized applications and get them to work together?

Chapter 8 >_Unicorns Are Real

- **Ben:** "I needed that software, so I built it, and then I open sourced it because other people needed it too."

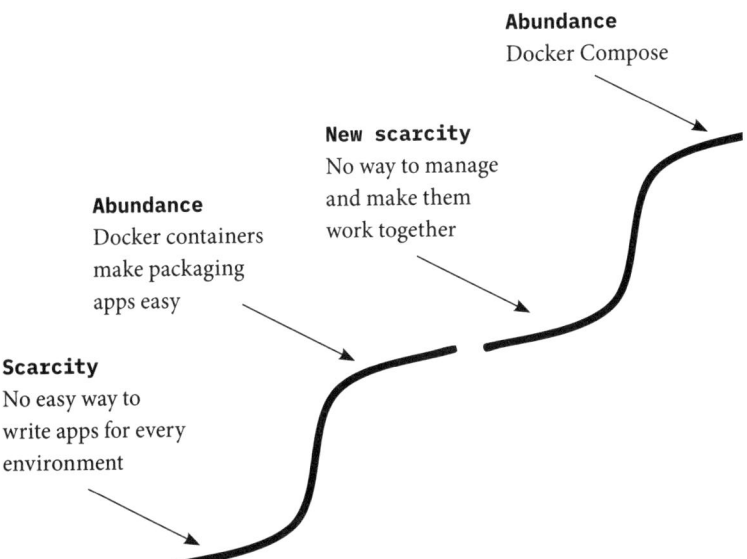

Every time tech fixes a problem, it creates a bigger one. It's not failure; it's the business model. Welcome to the S-curve, kid.

So how do you do it? Stay curious, and let that curiosity keep you at the leading edge of tech. Look for the abundances and learn to spot the scarcities. The more actively you are participating in the community, the more obvious the scarcities become.

S-curves aren't just how you start a company; they're how you keep it healthy in the long term. Great companies continue to feed the abundance of the first S-curve while executing on or acquiring the abundance of the second, and the third.

Why? Think about what happens when scarcities become abundant. What would happen if someone discovered a mountain made entirely of diamonds, and further investigation showed that the entire mountain range was full of diamonds? The price of diamonds would tank because any scarcity that becomes abundant loses value.

Let's pretend. It's 2021, and your startup sold 500K widgets for $20 each this past year, clocking in $10M in revenue. You all missed the S-curve memo (lots of startups do), but you didn't escape VC-driven growth demands. Let's play out your next few years using the table below.

Abundance in the industry is driving the price of widgets down by 20 percent each year. Thus, to continue growing by 30 percent, you have to sell 63 percent more widgets every year. That means your new revenue target is $13M, and so you have to sell 812.5K widgets instead of the measly 500K you sold the year before. Obviously, nobody needs to remind you that hot startups are doubling, often tripling revenue year over year, and you're growing at a mere 30 percent.

	2021	2022	2023	2024
Revenue with 30% Year-over-Year Growth	$10M	$13M	$17M	$22M
Price per Widget	$20	$16	$13	$10
Widgets Sold to Hit Revenue $	500,000	812,500	1,320,313	2,145,508
Growth Rates of Units Sold		63%	63%	63%

Ignore what I said before. You now know how to pick out a knife, sharpen it, and hold it. Let's chop some onions.

Chapter 8 >_Unicorns Are Real

Flywheels (AKA: How to Grow a Company)

Flywheels are a sacred metaphor of strategy teams that was popularized by Jim Collins in the book *Good to Great*. This concept was initially difficult for me to understand because I don't have a mechanical engineering degree and because I don't play with flywheels in my spare time. I get it now, though, and it's a beautiful metaphor.

Part 1: Momentum

Imagine, for the sake of this example, that a merry-go-round is our flywheel and you're still young enough to play on or around them without puking. You're meeting your friends Bob and Alice at the park, but you've arrived first. You approach the merry-go-round and give it a push. You then take a step back and the merry-go-round continues spinning for a few seconds. Interesting, right? It's storing energy when you push it and using that stored energy to continue spinning when you stop pushing. You approach the merry-go-round again and give it a push, and then another, and then another but begin to get tired. You step back and see that the merry-go-round keeps spinning for longer than it did with your single push because it has more stored energy.

Your friends Alice and Bob arrive at the park. The simple exchange of nods becomes a shared plan. The three of you approach the merry-go-round and Bob attempts to push the merry-go-round. Darn, he's at a weird angle and it doesn't go. Alice smiles and gives it a good push. Then you do. Because the merry-go-round already has momentum, Bob is able to give it a push also. Every push is easier than the last. The cycle continues, and the merry-go-round spins faster than ever. You each take a step back and the merry-go-round continues spinning and spinning and spinning.

That's when it hits you. Every push on the merry-go-round represents part of your startup strategy. You are a new product feature. Bob is a new user that wanted that feature. Alice is a new user base that heard Bob talk about that new feature at a conference. You realize, *It's not about maintaining speed—it's about accelerating it!*

Your heart is racing, but then the real wave of enlightenment hits. As if you've solved a mystery, you think, "Oh . . . *ohhhh . . . ohhhhhhhhhhh.*" You smile at the simplicity of it. The faster you can make your strategy flywheel spin, the easier it is to make it go even faster and faster. The phrase "exponential growth" has just entered your head when you see that Alice has waved a group of kids over.

Part 2: Self-Reinforcing Loop

The kids hop on as you, Bob, and Alice push. At first it's harder to push, but there's something about the extra weight from riders that causes the merry-go-round to go faster than before. You kind of remember something from school about more mass resulting in more momentum (this is the AP version of heavy things fall faster). You take a step back to ponder this. Bob and Alice also take a break. You marvel as the merry-go-round continues to spin and spin and spin because more mass means more momentum which means the merry-go-round has more stored energy! Your brain feels maxed out on this metaphor when a couple kids hop off, give the merry-go-round a few pushes, and hop back on.

By golly, if you build your strategy flywheel just right, it will become self-reinforcing.

Chapter 8 >_Unicorns Are Real

You, Bob, and Alice immediately apply to Y Combinator's Startup Accelerator.

>_

The most-often referenced flywheel in tech describes Amazon's growth strategy. Articles often get the story wrong, stating that Jeff Bezos sketched it out on a napkin in a coffee shop. The actual story is more interesting and is detailed in Brad Stone's book *The Everything Store: Jeff Bezos and the Age of Amazon*.

On a Saturday morning in early 2001, Jeff met with Costco's cofounder, Jim Sinegal. Jim explained that customer satisfaction was key to Costco's growth and their strategy was centered on this. Costco's low prices generated tremendous sales volume. That sales volume resulted in buying power for Costco, which they used to negotiate lower prices from their distributors. Ever-increasing sales with ever-improving margins that are achieved without raising prices on the end consumer is a solid strategy. You might even call it a virtuous cycle.

At a leadership offsite a few months later, Jeff invited Jim Collins, who was about to publish the book *Good to Great* and in doing so coin the concept of strategy flywheels, to present his research findings at the meeting. As part of the discussion, Jeff and his leadership team created their own version of this virtuous cycle/flywheel/self-reinforcing loop, which Brad Stone says went something like this:[16]

> Lower prices led to more customer visits. More customers increased the volume of sales and attracted more commission-paying third-party sellers to the site. That allowed Amazon to get more out of fixed costs like fulfillment centers and the servers needed to run the website. This greater efficiency then enabled them to lower prices further. Feed any part of this flywheel, they reasoned, and it should accelerate the loop.

Folks in the industry diagram it as shown in the following illustration. Some even do it on a napkin for added flare.

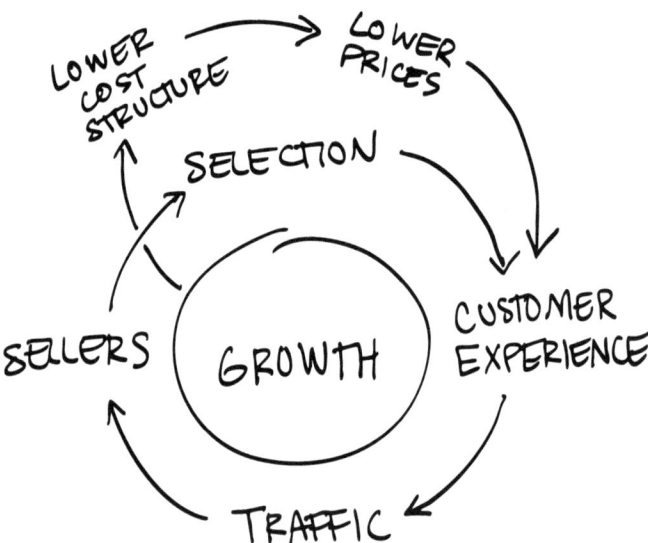

This is what it looks like when momentum becomes your cofounder. Build the right loop, and every small push from the team starts to carry the whole thing forward.

This flywheel is the strategy backing one of the largest tech companies in the world. Simple? Yes. Effective? Yes. Durable? Yes?

Chapter 8 >_Unicorns Are Real

S-curves and flywheels are the two biggest strategy tools you'll need. I do, however, highly recommend you take the time to deeply digest two *Harvard Business Review* articles that I refer to on a near weekly basis:

1. "Simple Rules for a Complex World," by Donald Sull and Kathleen M. Eisenhardt
2. "Bringing Science to the Art of Strategy," by A. G. Lafley, Roger L. Martin, Jan W. Rivkin, and Nicolaj Siggelkow

You just worked really hard, and now we've gotta do another hard thing. We've been talking about strategy at a very high level. Now, we need to get hands-on and build some shit. But you have worked hard, so here's a story as a palette cleanser.

My first job in tech was as a sales engineer. I'd never made that kind of money and so I bought myself some nice things. One of those nice things was a designer wool coat. I'm allergic to wool, so I couldn't wear the damn thing without a scarf. I flew down to Alabama to do an install on an Air Force base, and because I was going to be in a server room for two days (server rooms are cold), I brought my coat. Because I eat ADHD for breakfast, lunch, and dinner, I did not bring my scarf. That damn coat itched me so badly that I had to sit with my coat down around my shoulders all day. That afternoon, I bought myself a sweater so that I didn't have to wear the coat on day two. By the next day, it was just another story in my long list of ADHD mishaps, and I shared the story with the engineer I was working with that day and the one before.

He said, "Oh my gosh. I'm happily married and I thought you were hitting on me all day. You kept shimmying your coat down and I thought you were trying to show me skin or something." I was speechless. (I was also speechless a few hours later when we went to a Mexican restaurant for lunch. I choked on a pepper, excused myself, and puked it up in the parking lot.)

What lesson could he have taken away? "It's not about you!"

> _Chapter 9

Build Products That People Want

To go far, go together.

The way you do that is with a very detailed spec. Or what an engineer would call an API design doc. Or ops would call an NPI process. Or what marketing calls a launch plan. Or what a cartographer might call a map.

These are just different versions of creating shared narratives. Whether at a startup or a behemoth, the job to launch a product is to get everyone on the same page—before you build it. More people means more docs, more stakeholder alignment meetings, and more opportunities for things to go sideways.

No matter the size of the team, launching products is my favorite part of tech. The show draws me in, and the teamwork keeps me spinning. I love every bit of it.

My first life as an athlete taught me how to prepare for launch day. Decades of gymnastics performances and collegiate swimming races gave me the ability to endure months of preparation, navigate performance anxiety, and appreciate my teammates' execution. And just like every big event I've been through before, long nights, stage fright, and last-minute chaos always make an appearance. To this day, I still sleep terribly for two nights before a launch and forget to eat because my stomach is in knots.

I used to think that launching a product into the tech arena was all about the big day it went out. But after my first few launches, I learned the hard truth: Flying straight starts well before the launch. Just like every great work of art, like every major construction project, and every other thing in the world, success starts with solid planning.

Want to nail your product launch? You've got to start with the build. From my experience, there are three critical components you need to get right before launch day. Otherwise, you might be left wondering why Tinder beat out its rivals, why it takes four hundred pages to explain how to use some products, and why no one understands one another.

Phase 1: Write Down All Your Educated Wishes

Building and launching a product is basically a wild guess. It's your hypothesis about what others need, how often, and at what value. As Deadpool brilliantly put it, "I made an educated wish!"[17]

Before you dive into building, do your future self a favor and write down your guess. Yes, grab a pen, and define your product's hypothesis. Here's a simple template:

- I'm building [this product] so that [these people] can do [something better].
- I think this will work because [something you think you know].

Chapter 9 >_Build Products That People Want

- It might not work because [these people] also [like something else].

For example:

- "I'm building merry-go-rounds so that kids can play together."
- "I think this will work because there are kids who want to play."
- "It might not work because kids also like video games."

Boom! You just defined your product's hypothesis (or your supereducated wish). You've pinned down what's scarce (merry-go-rounds) within a population of abundance (kids who want to play) and flagged why it might not work (kids could play video games instead). You have your first launch hypothesis.

But before you pat yourself on the back, refine your guesses with data. Data keeps you and your team grounded as you build. The path ahead is full of opinions from General Mansplainer—aka, the loudest person in the room—and the only antidote is a data-backed educated wish. Here's our merry-go-round example with some market data and refined scope:

- "I'm building merry-go-rounds so that kids aged six to thirteen can play together."
- "I think this will work because there are a thousand kids aged six to thirteen within a five-kilometer radius of Spin City Park, and 75 percent say they want to play outside."
- "It might not work because 99 percent of kids aged six to thirteen also say they love video games."

These documented guesses will be your lifesavers after launch, when things might not go as planned. (Hint: They won't.) Even better, revisit and refine these guesses as you learn more about your market while you build your product and set up your flywheel. And as Southerners say, "Now we're cookin' with peanut oil!"

And if you think this sounds *so obvious*, let me tell you about Nuhook, my first startup during graduate school. Nuhook was an app for finding romantic partners. We obviously lost because you have no idea what in the hell a Nuhook is and are happy you will never see your last Tinder date again.

The problem with Nuhook's strategic direction was that we believed people wanted to find romantic partners only within their friend groups. We saw the abundance of people wanting to hook up, and we thought the scarcity was found in an app tied to your existing social circle. On Nuhook, your only options for swiping right were your friends from Facebook, Twitter, Instagram, and so on. Insider tip: Make sure that *someone* on your founding team is an extrovert so that you don't build an entire startup to get around a core issue—that you can't tell the guy across from you that you want to do very naughty things to him.

We might have seen the flaw in our thinking if we had written it out:

- We're building Nuhook so that people can find romantic partners within their friend group.
- We think this will work because people are already connected on Facebook, Twitter, and Instagram but don't have a safe way to ask a friend if they like them, ya know, in that way.

- It might not work because (1) people would rather hook up with strangers and avoid awkwardness if it all goes wrong and (2) lots of people are actually okay talking to other humans and don't need an app to talk to the person across the table.

The crux here is in our incorrect bet on how people make decisions. We were betting that the emotional awkwardness of asking someone out was worse than the awkwardness of seeing that person *afterward*. That statement alone says a lot about our experience because it turns out avoidance is the bigger motivator on the dating scene than being willing to step up to bat.

If we'd documented our educated wishes about Nuhook's abundance and scarcity bets, we might have avoided getting outsmarted and outswiped by Tinder.

Phase 2: Design and Build a New Ride

Now that you've got your educated wishes written down, it's time to design and build the ride.

This phase shifts from meeting-driven chaos to long, focused work blocks—the lifeblood of our overclocked misfits. Deep thinking happens here, where declining meetings and dodging interruptions isn't just a strategy, it's survival. One of my closest colleagues swears he needs a three-hour minimum to "get down into anything important." Sounds like the best vacation ever for my overclocked ADHD brain.

For early-stage founders, speed is the name of the game: fast iterations, rapid prototyping, and customer feedback on repeat. Your first job? Test your hypothesis before you run out of time, money, or patience. But once your product takes off, it's no longer just about moving fast. You have to keep the whole team aligned. That's when specs, defined roles, and shared direction become mission critical.

And since we're here, let's talk business models because without one, you're just burning cash for sport. Usage-based model? Engineering needs time to build telemetry without wrecking throughput. SaaS? Get multitenancy and security right, or you're in for a privacy nightmare.

Half of this phase is figuring out your flywheel. You should already know your customer, their problem, and how you're solving it. Now, stitch that into a customer journey and tattoo it anywhere you like. This is the backbone of everything you build and the tool to ground you when things don't fly straight. If your product isn't tied to a real need, then congrats: You just built a very expensive science project. Unless your business model is burning investor money for warmth, it's time to get real.

Let me say it only once: *It's the customer journey that's fueling the flywheel.*

Sketch it out, explain it in a doc, and get your fellow leaders and colleagues to comment on it. Stay there until your Driver agrees with the Amiable, and the Expressives and Analyticals align. Good luck. May your road map be clear and your stakeholders silent.

In my experience, it has never been the quality of the design spec that derails product development. It's the endless debates over something that can't yet be verified that burn through company cash. My god, I've seen so much time wasted over frivolous discussions that were not even going in a direction that could move the needle. Ever

Chapter 9 >_Build Products That People Want

say or hear "Let's punt that decision"? Look around one of those situations and you've found what we're talking about.

One of the most exhausting builder debates I've seen in tech was over query languages for graph databases. I could explain what that means and dive into the nuances, but who cares? The real issue was one guy not liking another guy's perspective, leading to two different solutions for the same problem. Neither worked as well as what we were already using, so customers went along with it with about as much enthusiasm as a toddler trying asparagus for the first time.

The answer to these situations? Show me the data! Measure what your customers actually use, and prioritize that. Use data in your meetings to avoid constant pivots and building only what the loudest person in the room wants.

Phase 3: Set Up Ops

This is one of those things that you think is for big companies and are wrong for thinking.

Once the ride's designed, it's time to make sure the whole thing can even run. Setting up operations is all about making sure the flywheel will spin and reporting on how well it does. And don't worry, the Analytical exec will be back in full force, cutting down your focus on revenue metrics and saying, "It looks pretty, but when you dissect the revenue, you see huge churn rates and most spending driven by curiosity."

Spoiler alert: He's right. That's the whole point of the flywheel—optimizing customer experience for patterns in product usage. Find the most trafficked paths and pave them with gold. That's how you reduce churn rates, find repeatable spending patterns, and fuel growth.

The key here? Operationalizing data, or corporate speak for making sure everyone uses numbers instead of gut feelings.

Ten years into my tech career, a CEO handed me the chief data officer hat. My first objective was to set up data operations around the whole company. As he put it, "I feel like I'm flying our plane blindfolded. Fix it." I'll tell you exactly how we fixed it in Chapter 13. What's important now is why we had to.

At the time, our product was running on gut feelings, with siloed teams telling completely different stories. Engineering bragged about open source activity, while marketing and field teams focused on enterprise customer journeys. Each team had its own definition of a "customer"—for engineering, a user didn't need to pay, but for the marketing and field teams, that detail mattered *a lot*. It felt like speaking five different languages just to figure out who had actually paid us a ton of money and was now complaining that the product wasn't working.

APIs, the digital handshake for sharing data, are the key to streamlining this chaotic process. Whether fully automated with code or structured as a workflow between teams, APIs help build a flywheel by ensuring the right teams are connected. Our data ops organization defined owners, identified the key metrics to share, and set deadlines for delivering that data in a specific format.

Teaching our entire company to find trust and value in data was a rough road, but the end result transitioned prioritization discussions from "Well, so-and-so said" to "The data says." And luckily, data doesn't ghost you after meetings. See, your team's data guru is

Chapter 9 >_Build Products That People Want

just an overclocked misfit who shuts you out when she goes heads down to build. You can't interrupt her all the time or the product will never get built.

So do yourself a favor. While engineering is building, think through the entire cycle of your flywheel. Make sure the customer usage metrics are ready to feed back into engineering, marketing, operations, finance, the people selling the product, those investing in the company, customer support, legal, IT, product management, and HR, because someone is definitely quitting over this launch. We will show you how to do all this in Chapter 12: Data Isn't People.

The punchline? As you make your rounds, align owners and metrics and thread data around your flywheel to keep it spinning. Because when the data spins, the money rolls in.

The Countdown to Launch

Congrats, you've made it through the build phase, where the real MVPs are born and the flywheels start spinning. But don't pop the champagne yet; we're just getting to the good part. Now it's time to launch this thing into the world, where everything will either go off without a hitch . . . or burn in a spectacular dumpster fire.

Either way, it's showtime.

>_Chapter 10
Stick the Landing

There's more to launching tech products than writing code. Every launch has a story. Something breaks, a feature or messaging misses the mark, or the whole thing flops. Maybe your CMO tweets a purple eggplant emoji at your CTO during his keynote *as encouragement* without knowing what it means.

Failures, or happy accidents, are inevitable. More often than not, these blunders have nothing to do with the source code. Learning to see them through the lens of scarcity, abundance, and flywheels is key to becoming a seasoned pro. So is keeping up with emoji meanings. 🍆💬👥♂🗒🚀💥

A launch brings together everything you've heard so far. The overclocked misfits, the opinionated leaders, and the unsung heroes all rally around addressing a scarcity, hoping to turn it into abundance and make money in the process.

Launching a product is more than just making your code public. You get one shot at a launch. Unlike Elon Musk, whose failures explode on a stage none of us will ever reach, the mistakes ahead in this chapter are the ones you're actually likely to make (we did) and recover from (we did).

Since we chased the Cro-Magnons out of the evolutionary tree, *Homo sapiens* have told stories to share life-or-death wisdom, teach family (tribal) history, entertain each other, instill knowledge of where the great berry patches are, and how to stave off the other tribes from attacking your settlement. Eventually, we innovated how to write them down. We are wired to learn from stories and bits and pieces of stories. It remains true in our technotopia, or technodystopia, that words matter just as much as code.

Because a great product without the right story is just another code repository lost in the void.

Positioning Your Product: Words Is Important

Remember those educated wishes? Since then, you've been building a product and solving a real-world problem with brilliant engineering. Now, the tech is airtight, the features are impressive, and the team is hyped.

Messaging is where most teams mess this up: They focus on what they built instead of why customers urgently need it.

You built this thing because you found a scarcity that needed to be addressed. Now, as you introduce it to the market, your job is to make sure people clearly understand that you're solving *their* underserved needs. The way you do it—through the words you choose and the stories you tell—lets customers realize for themselves that your product is exactly what they've been looking for.

Want to nail it? Use their words, not yours.

There are a couple paths you can take.

Chapter 10 >_Stick the Landing

Vitamins Versus Painkillers

What's the difference between a painkiller and a vitamin?

A painkiller solves an urgent, high-stakes problem. Customers are actively looking for relief. They don't need convincing—they just need to know where to buy it. A vitamin, on the other hand, is a nice-to-have. It might improve life in the long run, but no one is waking up at 3 a.m. in a cold sweat over it. The difference in how customers perceive a product, whether it is a must-have or optional, determines whether your launch soars or sinks.

The Hero's Journey

When you are determining the positioning of your product, you need to think like a storyteller (not a builder). A tried-and-true recipe for this is the hero's journey.

Here's a hero's journey for a techie: Imagine an overclocked misfit grinding away, dreaming of greatness, but stuck in a loop of fixing bugs and dodging Slack pings. Then, catastrophe strikes, code breaks, deadlines implode, and they're one purple eggplant emoji away from losing it. But they find a secret tool (probably some game-changing framework or AI), and just like that, they push to production, the flywheel spins, and suddenly they're the hero who saved the world with a git commit (our way of saying the hero stuck the landing).

Successful positioning frames your product as the hero's tool in your customer's journey. That is how you make your product a painkiller instead of a vitamin.

Talk to your early customers. Find out how they see themselves in the hero's journey and try to empathize with their experience. We call this *outside-in thinking*.

Reframing is hard, and the closer you are to building your product, the harder this is. Telling your product's story is an exercise in stepping outside your own experience and truly empathizing with your customer's challenges. But this is the good kind of hard work. It's the kind that builds shared empathy and makes your launch story compelling. When you get this right, your message will resonate, build trust, and help people see themselves in what you've built.

As someone who came to tech from the engineering side, learning to think from the *outside in*—through the customer's eyes—was one of the hardest adjustments to make. It's disorienting to let go of precision, technical correctness, and the urge to explain how something works in favor of speaking to why it matters.

Every successful engineering innovation goes through this transition eventually. A product isn't successful because it's built well; it's successful because people understand *how it helps them*. The shift from "This is a brilliant solution" to "This solves your problem" is what turns an invisible technical breakthrough into something people care about.

And that's the whole reason we build in the first place.

It Costs How Much?

At this point, you've nailed the positioning, the story is compelling, and your product finally sounds like a painkiller instead of a vitamin. People get it. They *need* it. And now comes the moment of truth.

Chapter 10 >_Stick the Landing

"Okay, but . . . how much?"

This is when engineering looks at you like you just asked them to put a price tag on their firstborn. And, don't forget that your Driver exec expects you to magically conjure up a pricing model that makes sense, drives adoption, and doesn't tank the business.

Welcome to the fine art of pricing, where what something costs and what people think it's worth are two completely different beasts.

If I'm asking engineering to sit back on the market story, pricing is where they're going to tell me to sit back down. You think you're pricing a product as you build it, but I've only ever seen it done backward. Engineering hands you a pain pill, and now you're the one stuck figuring out how to tell the world what it costs.

Do your future self a favor and think about pricing while you're still building. If you need to charge for every read and write, you better know how many of those are going to happen. Telemetry impacts product performance, and engineers won't want anything clogging up their precious throughput.

This is a numbers game. You're dealing with leaders who are *Analyticals*. They care about accuracy. Your job is to figure out exactly how much this new product costs. Work closely with engineering to calculate it. Think through how the cost changes from one to ten customers, then from one thousand to ten thousand. While you are there, find where your infrastructure is wasteful and put it on your backlog until bleeding is a problem. Open a new tab in your spreadsheet and start running some numbers.

Pricing is all about margins. The amount customers pay above your operational costs is what funds your company's growth. We *could* tell you what margin to aim for, but, well . . . you're the competition now.

Next, you've got to figure out how to *sell* it. This is why MBAs get paid the big bucks. This is the art of combining the hundred-and-twenty-eighth revision of your marketing words with the mechanics of your product's flywheel. You need to set a price that gets enough people hopping on and paving paths through it. You need to think through how pricing adjusts at scale. What's the smallest bit? The largest? What increments do customers expect in between? Is your smallest option cheap enough for your customer to test drive, or have you already priced them out?

All of that has to make sense *and* form a margin, or a path to an acceptable margin, that your execs will sign off on. The tension in the room around this conversation is inversely proportional to the size of your company.

Now that pricing's settled, it's time for the next boss battle: naming the damn thing. Have marketing bring snacks because you're about to spend two weeks arguing over whether *Flow* or *Stream* sounds cooler. Agreeing on that will take more energy than it did to build the product.

What Are We Going to Call It?

Naming a product is a lot like naming a kid. I want to be in charge of doing it. I want it to be cool. And, I want others to wish they'd thought of it.

Naming a flagship product is different from naming a secondary product or feature which is either free with or charged as an add-on to the flagship product. If you're first to market, things are

Chapter 10 >_Stick the Landing

easier. Unless you're aiming to be the next Xerox, consider choosing something that reflects the vision of what the product does or what people are looking for when they find it. Adobe sets the bar on this. The name Photoshop captured the vision of the software and has been successful enough to grab secondary meaning as the action people take when they edit photos, in the same way Xerox did in becoming the action of copying files ("Can you xerox that for me?"). Those kinds of trademark problems are something we all dream of having. When Adobe launched Creative Cloud, it wasn't cloud anything. But the vision was to be a collection of services, with the artist's work accessible via a range of products. The name matched the vision and the vision became reality.

If you're not first to market or if you're launching a product within an early market, it's tempting to latch on to the name of seminal companies. Remember the iEverythings of the 2010s? The LangX's of the GenAI wave? These naming choices might get your product found, but you're not building your own brand equity. It's also a questionable strategy to name your product after something that is popular in the first minute of the first inning of a technology wave.

You should also avoid the overly cutesy name that enterprise architects cringe saying and that line of business owners question the validity of. SurveyMonkey is incredibly useful, but is it something you buy an enterprise contract for?

That brings us to changing company or product names. This is rarely, if ever, worth the time and money. The exception is when you're not changing the name but launching a new product that is almost exactly the same as an existing product. This can be a great strategy. A lesson from *How to Win Friends and Influence People* is that humans *love* things named after them. They also love things made for them. Salesforce's Health Cloud and Financial Services Cloud told every buyer in those industries that something existed just for them because they're special.

Lastly, times change. If the name you choose has any association with a single group of people, abandon that name. It doesn't matter if you didn't know about the association, if you can talk your way around it, or if you didn't even know about it before someone mentioned it. Abandon that name.

If you are naming a feature, name it based on what it does. If it adds charts to your flagship product, it's called Flagship Charts. If it adds ears, you can call it Flagship Hearing, but that's as creative as you should get. You want existing users to want it. You want new users to know you have it and others don't. Do not confuse the users and don't anger them. If the new feature does adding and subtraction, don't call it Flagship Analytics. Features don't get as much leeway for "vision" names. Name the feature what it does.

It's okay to break most of these rules. I have. Some folks will adamantly disagree with my guidance. I once stood speechless while an exec stepped through all of the reasons that one of our products was "by far the worst product name" he'd ever encountered. It latched on to an industry trend and wasn't something users would understand at first glance. I laughed, admitting that I'd named it. He looked me in the eyes and said, "It's really terrible." Again, naming products is like naming kids: With a little effort, you can get another chance. That exec loved the next product I named, trademark issues and all.

Chapter 10 >_Stick the Landing

A marketing exec was who shared the Adobe stories with me. He was at Adobe when they chose Creative Cloud and pushed back on it because he was only just about to learn the vision versus reality rule. Remember, unless the name aged like milk, a bad name makes a good story.

Go for a Test Ride

You aren't just launching a product—you're launching a flywheel. This is your first push of the merry-go-round, and if you didn't check the bolts before inviting users to hop on, don't be surprised when they're flung off at full speed.

When ops is scheduling, ask them to pick an ideal date and then back up two weeks. Budget more time than you think because the build will be late, and somewhere along the way, engineering started believing their deadline was launch day. Surprise, code freeze was supposed to happen a month ago, but somebody slipped in a new feature that we didn't plan for. Nothing like an eleventh-hour Hail Mary to keep things interesting.

History is littered with great products that failed spectacularly because someone skipped this step. Tesla's "shatterproof" Cybertruck windows shattered twice during its live demo.[18,19] Apple's iPhone 4 antenna was a masterpiece of design until people actually held the phone and realized their hands blocked the signal.[20] Samsung spent years hyping The Wall, a futuristic modular TV, and then nobody could actually figure out how to install one.[21] These weren't bad products. They were just products that didn't survive reality.

Failure comes in all forms, but the worst ones follow a pattern: overpromising and underdelivering. The big feature isn't done, so the product doesn't deliver on the marketing promise. Performance, security, or throughput metrics are wrong, so users can't rely on it. Or worse, the product experience is so bad that even if it technically works, no one ever wants to use it again.

MVP stands for minimum viable product, but people forget the *viable* part. If your first reviews say it barely functions, you've already lost. So, test your product. Then test it again. Because no one wants to be the one rewriting the launch blog post to explain why "unexpected behaviors" are actually features.

Launch Day

On the biggest launch day of my career, it wasn't my 5:30 a.m. alarm that woke me from four hours of sleep; it was my pager at 4:13 a.m., signaling trouble in the final hours of deployment for the mainstage. Numb to fires after a day of pivots, I answered this one with the apathy of an athlete on day seven of a seven-day tournament: exhausted, but ready for the next battle.

The engineering lead informed me the pipeline was stuck. But we needed to launch and land our service around the world within a small window of time that was approaching fast. I made calls, got approvals, and helped clear the blockage. Another fire was extinguished. Time to head to the launch room under the mainstage and stick the landing. Going back to sleep was out of the question at that point. So, I decided I might as well enjoy a sunrise walk with my favorite warm beverage, snap some selfies with the Sphere, and enjoy a slow pace while I could.

Chapter 10 >_Stick the Landing

Aside from that one early-morning page, the rest of the day went smoothly. The launch room was like mission control. I reported to the lead operator, set up three launch checklists across my screens, and spent hours in that basement, my only words being, "We're a go."

Our mainstage moment came and went. I clicked boxes on a spreadsheet, and it was over. The buildup for the launch was finally done. I felt at peace.

That's the beauty of working with world-class ops people. They get things done calmly and efficiently, and the overclocked product lead doesn't fall over from panic. No one outside that room knew the criticality of what was happening in there. Jumpstarting new flywheels is an ops person's love language. If your launch days don't go smoothly, there's your first place to look.

And that's product life, folks. Too much anticipation for a moment that feels forgettable, yet is anything but. I'd almost prefer the numbness in my legs as I kick through the last lap and hit the wall of the pool, gasping for air. At least then, you know right away if you won or lost the race.

Keep Pushing

After launching a product, as the general manager of my service told me in that basement, "It's like being in the eye of a hurricane: calm, sunny, but with an eerie acceptance that another hit is coming."

So what comes after the launch? Welcome to the next big chapter in development: creating and establishing growth. This is where your product becomes your best marketing tool.

Ever forward. 👩‍💻💻🚜

>_Chapter 11

Grow the Flow

What happens next is largely situational. If you're at an early-stage startup, this is your situation:

None of your initial hypotheses or wish-list items will survive contact with the market. You must now find repeatability and then optimize for it.

Product launches are often seen as the finish line. When things get tough, we say, "Let's just get this over the finish line." The word "just" implies that the finish line is not a finish line. The word "just" is offered as a safety net that we interpret to mean, "And then we will . . ." What? Finish it again? Yes. Exactly.

I ran a product-led growth team before understanding the importance of repeatability. I had no idea I was set up for failure. I was putting every hour of my life into doing everything described in this chapter, and it wasn't paying off. I assumed that the project was failing because of my lack of experience and because of a constant lack of resources. I also couldn't understand why the product and engineering teams wouldn't prioritize the fixes we needed based on the friction users were experiencing.

Looking back, I see that I was looking for optimizations based on ten or more user journeys I thought we needed to support. We didn't have something that could achieve repeatability *yet*. If every user that you talk to describes a different type of friction, there is a bigger issue to address before you take on product-led growth.

Months into the effort, I opened a book on product-led growth, and the first sentence was something like, "You can't grow a product that people don't want." I knew we didn't have product-market fit. I closed the book. We didn't need it yet.

Find Repeatability

Repeatability means that people are adopting your product for the same three to five reasons and achieving the same outcomes over and over again.

If users are buying for more than three to five reasons, you're selling edge cases. You are in a situation where your headcount isn't going to scale with your growth because virtuous cycles need repeatability that you don't have. This is bad. To succeed, you must constrain your value prop. You have to be willing to tell a customer "We don't do that" and lose the deal. If you are telling a potential customer that your burger place can also make pizzas, you're selling an elasticity that doesn't exist.

Think back to our flywheels. What should you do if a flywheel isn't spinning? You have to fix the flywheel. Growth is about making something spin faster.

There is some aspect or capability of the product that users like. Iterate on that functionality to grow its value toward repeatability. You'll have to block the noise. There is no such thing as table stakes at this stage of the game. You do not need x security feature or z latency. You need something people want.

Get Pull or Pivot

The right market pulls a good product. If your product isn't getting pull, it's probably in the wrong market. You need to pivot. You may need to pivot a lot. This is something the Y Combinator Startup School graduates get right. They know that the startup game is a lot like poker. If you fold early and fold often, you don't lose your shirt.

A strategist once advised me that "if the engine starts going sideways, let it." Users might like a product for a variety of reasons. You must find the reason or positioning that offers differentiation from the rest of the market. Here it is again: Care about *their* words and not yours.

Map It. Measure It.

Welcome to the party! We're so happy you took however much time it took to get here. You know your user and you know why they're buying your product. It's time to draw out the user journey.

User journeys are often drawn as funnels because getting a purchase order or credit-card swipe is usually considered the win. You know from the flywheels chapter that getting that paying user to bring more users to the product is the real win. We're fine with either flywheels or funnels for this use case.

Regardless of how it's drawn, to evolve the product and grow your business, you need to do three things. All of them are going to take absolute maniacal focus.

> 1. **Know the flow:** Understand the user journey (flow) and collect metrics that show how each user is progressing in their journey.
> 2. **Observe the flow:** Uplevel product telemetry to serve as the source of truth for tracking adoption trends and identifying issues. This is where you'll make and evaluate experiments.
> 3. **Grow the flow:** Make product decisions based on data-driven observations of user activity.

Flow refers to how users flow through the funnel or flywheel. Knowing the flow means understanding the user journey and collecting metrics that show how each user is progressing in their journey.

For each user or enterprise, there is a rough journey from when they search for a solution, find your solution, use it, use it more, make a purchase, and then recommend the platform to others. You must be able to measure each of these steps ideally at an individual user level, but at minimum you must be able to measure their rates as a whole.

So how will users find your website? Are you the vendor behind a popular open source project? If you aren't, you should be, but that's another whole book. Are you running ads? Are you hustling by answering Stack Overflow questions and slyly pointing to an example on your website? Are you conducting workshops and advertising

Chapter 11 >_Grow the Flow

them with social media ads? Are you publishing tutorials on sites popular with developers? These are all top-of-funnel efforts, and some percentage of folks that do them should go on to do something else. A workshop attendee's next step might be to register for your service. The right users will take an action on the service and then take additional steps that show more use, which becomes more consistent and leads to a purchase. Pave this path with gold.

Every step is a metric, and those metrics should all go to a system of record before being upleveled to engineering, finance, marketing, and product. This is important: All metrics should go to the same place. Nobody skips the system of record because data isn't for one person or one team. Hoarding data or siloing data will limit your growth. It should be shared generously but responsibly.

The next step is to make the data observable. This means it needs to be usable by downstream teams using the tools of their choice.

The user journey should be broken down into stages, and you should have absolute numbers for the number of folks entering stages, but it's more important to focus on the conversion rates between stages. The initial focus of growing a product with product-market fit is removing any friction that keeps users from converting between the stages. This is the difference between celebrating a 100 percent increase in ad clicks and realizing that the bigger win would have been a 2 percent increase in conversions from ad clicks to sign-ups. This is also where you learn that chasing adoption metrics is gameable, and there are a lot of reasons to game it.

An artificial boost is akin to a really big initial push on the merry-go-round. If you're going after funding, traction metrics matter. Gaming metrics has its limits. You can buy GitHub stars, page views, and sign-ups, but investors will expect you to keep that growth rate up. And ads cost money. If the flywheel isn't taking off, you'll go broke faking growth. You also won't be able to understand your growth formula. I listen to competitor earnings calls like our livelihood depends on it. "Because of the economy, expansions are slowing, so we're pushing to land more new logos." That's a company that knows their expected expansion rates. They know that if expansion is slowing by x percent, they'll need to make up for that loss in revenue by adding y percent new customers. You can't do this math if you're gaming metrics every quarter.

To grow the flow is to take action on the insights you observe and to remove any friction that is keeping your merry-go-round from spinning faster.

Growing the flow requires an experimentation framework. Take the ad clicks example. What could have caused a 100 percent increase in ad clicks without impacting conversions to sign-ups? Running ads for the keyword REST API when you sell a database will result in a lot of clicks but very few conversions (clicks that become users). An experimentation framework lets us test and retest hypotheses. Put all ideas in the experimentation framework (it's a spreadsheet, but experimentation framework sounds much cooler). Decide which experiment to do first. Identify the success metrics. Do it. Measure it. Iterate on it or move to the next hypothesis.

Here's a snapshot of our experiment tracker. Each row shows the idea, the goal, and the outcome. In this one, we changed the sender address from "hello@" to a real name and lifted our welcome email open rate by 7 percent.

Chapter 11 >_Grow the Flow

Experiment Type	A/B
Category	Email
Experiment Description	Welcome Email Open Rate Test
Hypothesis	Users are more likely to open an email from a person vs. a generic company email address.
Experiment	"A: 50% of welcome emails sent from hello@company.com B: 50% of welcome emails sent from realfirst.last@company.com"
Success Metric	Open Rate
Tool	customer.io
Status	Completed
Results	"A: 50% open rate B: 57% open rate Experiment complete: Welcome email updated to B"
Effort	Low
Owner	Alex
Outcome	"Success. 7% increase in open rate. Welcome emails will be sent from an identity"

The team's goal is to remove friction and make the flywheel spin faster. Pricing confusing? Add a calculator. Project creation screen confusing? Redo it. Is the examples screen getting tons of clicks, but no examples are launched? Talk to the users. Ask what they'd hoped to see. Deliver the examples they wanted.

Making a flywheel spin is addictive. Pause to celebrate the team's successes. Custom swag, like belt buckles, is always the best way to celebrate a team's success.

These are the kind of people who turn a spreadsheet into a strategy, a user interview into a road map, and a swag belt into a badge of honor.

>_Chapter 12

Data Isn't People

Back in Chapter 9, I shared how a CEO handed me the chief data officer hat and said, "I feel like I'm flying this plane blindfolded. Fix it." At the time, our company's data operations were barely more than numbers on slides. Our leadership meetings were Monday-morning accounting exercises. Everyone was building products and features, but nobody was monitoring how our big bets were playing out. We had marketing and sales chasing leads, engineering juggling features, and finance tracking budgets but no shared narrative on whether we were actually on course or heading for a nosedive.

Six months later, we built what I liked to call our "cockpit." Every Monday, the leadership team got together and reviewed a set of data-driven spreadsheets. (Because every company is run on spreadsheets.) We started Mondays with real data that showed us where we stood, what was slipping, and what was just plain broken. We started the week from a shared place of trust. It was no longer about whether the numbers were perfect; it was about how to use them to our advantage.

On a longer cadence, we also reviewed the newest industry data to see if our trajectory matched what the biggest enterprises were doing, or if we were way off. We valued that industry trends were directionally correct. Our secret sauce was creating and testing shared hypotheses about where our company fit into the tide.

It wasn't a dramatic makeover. We simply started using data to inform our decisions in real time. Now, our Monday meetings weren't about defending whose numbers were right; they were about using the numbers to plan what's next. When we saw that new sales reps were taking months to ramp up, we drastically changed our onboarding process. It might not be glamorous, but it worked.

Anytime that CEO was asked how some aspect of the company had been turned around, he was able to provide a data-driven answer. The process provided him with a log of decisions backed by data and reasons for change. That is how you build trust and drive companies to exist with data, not egos.

Here are the six things we did, and here's why we recommend them for leaders who are ready to move from chaos to control.

1. Data Is the Grounding of a Shared Narrative

It's never really about the data itself. It's about creating a framework for conversations, for making guesses, bets, and hypotheses about what might happen next. Data gives you a shared language. Without it, your meetings are just ego-fueled debates, and the loudest person in the room usually wins.

I've been on engineering teams that prided themselves on being "customer obsessed." But when I went around asking forty different leaders what data they needed to make better decisions, they all said the same thing: They had no idea what their customers wanted

Chapter 12 >_Data Isn't People

because there was no data to guide them. Meanwhile, engineering kept pushing out features no one asked for. Without data, they were basically throwing darts blindfolded, hoping one would land on the bullseye.

Sure, numbers can be skewed. But if you're not even talking about the numbers, you're probably making decisions based on someone's ego or the loudest customer. So yeah, it's not about the data. It's about the conversations and the accountability that data brings to the table. Because nothing's worse than an ego driving a product road map.

2. Watch for Deadline Heroics

Let's talk about deadline heroics. Here's the thing: You're going to set goals, and sometimes those goals will be missed. What you need to watch out for are the last-minute scrambles where someone suddenly manages to "save the day."

For my first two years at a startup, our sales numbers were always in the red—until the last two weeks of the quarter. Then suddenly, everyone was pulling in deals at the last second, slamming into the deadline to bring money into the company. It wasn't because customers magically decided to buy all at once. It was because every deal was held together by last-minute heroics. Sales reps would panic, execs would get hands-on with customers, and we'd all scramble to convince them to believe in our product just long enough to close.

The first major change our new CEO made was to bring repeatability to the sales process so that money would flow into the company steadily. The shift for leadership was moving from firefighting to relationship building. Instead of swooping in at the last minute to fix operational details and relying on hope alone to close deals, leadership began building long-term, meaningful relationships with customers. They talked about vision, about where the product was going, about the future of the company.

If you're an executive who keeps getting pulled into the weeds with sales, take a step back and ask yourself what kind of conversations you're having. Are you constantly convincing people at the last second to sign? Or are you building relationships so that customers trust you long before it's time to renew? Because if every deal depends on last-minute heroics, all you're doing is training your customers to only buy when you beg.

3. Create a Drumbeat

Imagine data as a steady rhythm that keeps your team in sync. It's like a drumbeat that sets the pace for everyone. There's no rushing, no last-minute improvisations, just a solid groove. You don't want frantic bursts of activity that burn everyone out. You want steady, reliable progress.

In a restaurant, it's like the rhythm between the kitchen and the front of the house. The kitchen can't just cook at its own pace, ignoring the flow of orders. They need to listen to the beat coming from the waitstaff, adjusting the pace so the dishes come out just right.

Your team meetings should be the same. Organize weekly, predictable check-ins where you look at the numbers and see if you're still on track. But here's the thing: It's not just about stating a number and moving on. It's a discussion. Ask, "Why is this number higher or lower than last week? What did we change? Do we need to change more, or do we need to revert?" That's how you replace gut-feeling

Chapter 12 >_Data Isn't People

prioritization with a little bit of science. It's not flashy, but it keeps things running smoothly.

We've talked a lot about telemetry for product use, but data is a shared narrative across the org. Sales, marketing, engineering sprints, uptime, downtime, margins—all of it comes together to paint a picture of where you are and where you're going. Data isn't just a tool for measuring success in one part of the business. It informs leaders across the board, helping them make decisions that keep the entire operation moving forward.

You know you've got this right when progress is incremental, even boring. It's not about big wins; it's about keeping the ship steady and on course. Because if you're only relying on those last-minute heroics, you're heading for a burnout. I promise, that's not a good look.

4. You Can Change It

And here's the kicker: You can change it. You can change the metrics. If something isn't working, if the data is showing that you're not on pace to hit your goals, change the goal. Adjust. Pivot. Don't stick with a plan that's clearly leading you off a cliff just because it's what you wrote down three months ago.

The point of using data isn't to lock yourself into a rigid plan. Data is also the thing that tells you turbulence is ahead, so adjust as you go. Think of your plan like piloting a plane. If your instruments are telling you that severe weather is ahead, you don't just keep flying straight into it. You find smoother air.

Same goes for your business. If you're hitting rough patches, adjust the numbers and goals, and find a better path forward.

This is why I insist on tracking metrics. It's not about checking a box; it's about knowing when to pivot. It's about creating a culture where change is a sign of growth, not failure. It is also how you keep egos in check: by showing that the data supports the changes you're making, and that it's all part of the plan.

5. Things Get Worse Before They Get Better

Let's talk about one of the toughest parts of using data: Things often get worse before they get better. I learned this lesson firsthand. At the time, our developer workshops were pulling in big numbers: two thousand attendees and three thousand registrants. It looked good on paper, but we knew that if we wanted to scale, we had to shift our audience and change up the workshop content.

So, we made the change. And guess what? The first few workshops brought in a measly five hundred developers. We could have easily taken one look at that drop and said, "This was a terrible decision." But we didn't. Why? Because we knew that we were aiming for a new audience, and the initial drop was just a sign that we hadn't figured out the right recipe yet. We were testing new ingredients and refining the mix, and those early numbers were just part of the learning curve.

When you're using numbers to make decisions, you've got to be ready for things to look ugly at first. It's like running a restaurant that's switching up its menu. Maybe you've had a signature burger that kept the place packed, but you know that the neighborhood is changing, and people are looking for something fresher, trendier. So, you take the burger off the menu and replace it with some new dishes. The first few weeks, your regulars might walk out grumbling

Chapter 12 >_Data Isn't People

because they miss the burger, and sales might dip. But if you believe that the new menu is right for where you're heading, you've got to stick with it.

Just like our workshops, it's about understanding that those initial low numbers don't mean failure. Drops in metrics meant you were still perfecting the process. The worst thing you can do is panic, go back to the old menu, and try to win back the customers who are already eating somewhere else. Change takes time, even if the early signs make you sweat.

6. Two Types of Data, One Way to Use Them

And here's the practical side of things: There are two types of data you need to think about, and one way to use them. You've got operational data: the stuff that tells you what's happening day to day. And then there's strategic data: the information that helps you see where you're headed in the long run. The way you use them is *together*.

Let's go back to our restaurant example. Operational data is the daily stuff, like how much lettuce you're going through, how many tables you're turning over, and whether the dishwasher is up and running. It's essential, but it's not the whole picture. Strategic data is like knowing what food trends are coming next, whether customers are moving toward plant-based dishes, or if nonalcoholic cocktails are about to be the next big thing.

You can't have one without the other. If you're only focused on the daily grind, you might end up building something that's perfect for a tiny group of users but fails to grow. It's like designing a fancy tasting menu that only a handful of diners appreciate while ignoring the broader market that's hungry for comfort food.

On the flip side, if you're only looking at strategic data, you're just chasing trends and dreaming big. You will lose sight of what's actually happening on the ground. It's like a chef who's obsessed with creating the next big culinary trend but forgets to keep the kitchen stocked with basics like salt and olive oil.

You need both to keep your business moving forward.

ADDING IT ALL TOGETHER

Data is about much more than just numbers on a page. It's about creating a framework where your team can have real conversations without getting bogged down by egos. It's about recognizing when someone's pulling deadline heroics and knowing that's a red flag. It's about creating a steady drumbeat that keeps everyone in sync and knowing when to change the beat when the rhythm isn't working. And most of all, it's about blending operational and strategic data so that you're making decisions that serve both today and tomorrow.

I get that data can be manipulated. But you know what? It's better to have a conversation about why someone is tweaking a number than to have no conversation at all. At least then you've got something to work with.

Chapter 12 >_Data Isn't People

So, founders: Use data not because it's perfect, but because it gives you a starting point. Data gives you a way to keep everyone moving in the same direction. It's your best shot at building a company that's guided by more than just gut feelings and the loudest voice in the room.

>_Level 4:

It's a Jungle Out There

> 13: Enemies, Friends, and Frenemies ... As If

> 14: Get That Funny Money, Honey

> 15: Piercing the Dark Veil of Acquisitions

By now, you're probably used to your code base being held together with duct tape and good intentions, but there's a new twist in our tech arena: Framings change as you play the game at increasingly higher levels.

This is when the CEO of a competitor might become a confidant. You might be swapping investor leads with her over decaf oat-milk lattes at breakfast and then shaking your fist at the company she leads three hours later on an all-hands.

You're playing a broader game on a much larger playing field. Every relationship you make along the way now matters, and real competition is actually as rare as well-documented code.

You're also entering a new arena that has new rules. You must learn these rules to secure the capital you need without trading away the soul of your startup.

While landing funding gives you the runway to execute on a strategy, acquisitions, when handled correctly, are an opportunity to accelerate your strategy. It's kind of like using a whistle in *Super Mario Brothers 3*, enabling you to warp worlds ahead in the game.

And lastly, you must learn that there's always a chance that your princess is in another castle.

>_Chapter 13

Enemies, Friends, and Frenemies ... As If

I once listened to a podcast where a doctor that focused on infectious diseases described why she had the best job in medicine. She said something like, "This is where the juicy stories are." You know she's right. We've all lied to our doctors, but if you're deathly ill from a mysterious illness, the truth is going to come out, and when it does, Dr. Infectious Diseases will be there to collect that information (story) and, hopefully, deliver a cure.

I remember feeling so *seen* as she confessed the guilty pleasure she derives from her life's work. If Dr. Infectious Diseases is like me, she was called nosy as a child, was accused of gossiping as a teen, and learned the difference between gossip and intel by the time she was an adult. Gossip is gathered for its entertainment value and used for clout. Intelligence is gathered for its strategic value and used to understand industry trends, shape strategies, and even act as an early warning system of sorts.

There are times when, even as a tech industry intelligence specialist, I have fallen into the gossip trap. Is there a company you hate? Maybe a bunch of great employees left to go there. Maybe you suspect that their code base has overlap with your intellectual property. Here's the thing: You can't personify a company. I learned this while pontificating on the sleazy, dirty nature of a competitor that wasn't actually a competitor. I expected my coworker to join me in dragging those scoundrels through the mud, but instead he said, "Are you actually personifying a company?" This was one of many moments in my life where something I said very seriously in one breath felt very silly in the next. He was right. As a general rule, you shouldn't waste time hating something that can't hate you back. Your time is better spent understanding why those employees left your company (not you—your company) for the other one. Doing *that* is valuing information for its strategic value.

The way to do that is to respect yourself and other humans. Gossip ruins lives and careers. We often mask gossip as venting. Venting is tricky. The original idea for this book was venting adjacent, but you live and learn, and one thing we've learned is that venting is often our ego's way of avoiding self-reflection. Thus, it's something we can let go of.

There's a way to get ahead of gossip: It's to not get pissed off. The way to do that is to always *assume noble intent.*

- **Assume** conversations you engage in will be meaningful and substantive, even if it's uncomfortable.
- **Noble** means each party is representing a meaningful point of view that aligns with strategic priorities. This is where respect comes in.

Chapter 13 >_Enemies, Friends, and Frenemies...As If

- **Intent** refers to a person's higher-level purpose, beyond just their words and actions.

With an ability to maintain good relationships and a thirst for information, you'll begin to set aside other shallow conclusions you've drawn, like who is and isn't the competition. It's easy to fall into the trap of seeing other companies as competition when there's overlap on secondary feature sets of your product. Countless times I've been guilty of and watched others squabble over french fries when one company sells burgers and the other sells chicken nuggets.

I was provided a framework for identifying the real competition. To be a competitor, all three of the following questions must be answered with a yes:

- Do you serve the same user?
- Do you serve the same core use cases?
- Do you share the same buyer?

If all three answers are "yes," we're talking about a competitor, and there's a lot of information you should be interested in, but there's one more misconception we need to let go of. It's *not* good if solid competitors are doing badly. Pick any three to five similar public companies and overlay their stock trends. You'll see a rise and fall that's usually more indicative of industry trends than any one company. Publiccomps.com is a great resource for this. Create an account, log in, and check out the high-growth SaaS category.

Tracking and comparing metrics of your public competitors (or with whatever intel you can gather from friends about private companies) is essential. If you're competing in a new market and there are no public comps, you'll only be able to get so much information from friends, so you need to be really smart with your questions. Knowing their margins might settle a dispute between you and the CFO, but that's a penny-wise (ha) use of your questions. Is their product road map going to cut into your value proposition? That's meaty intel that affirms your strategy and also qualifies as an early warning system. If the competition makes a pivot, they might be seeing something about the industry that you haven't. That's information you need.

As a leader, you must learn to appreciate the competition. If you're the CEO, you should have the phone number of the competition's CEO. This is C-level inside baseball. To the rest of your organization, it's *highly* motivating to have an enemy you want to crush, but remember your manners. Assume that any competitive guidance you provide will be handed directly from someone on your sales team to a customer who will hand it directly to the competition. Now you've put your inside baseball game at risk. Be smart. Because to quote an African proverb, or an Anglo-Saxon proverb, or Hillary Clinton, or Cory Booker, or Rudyard Kipling (you get the idea—nobody knows who said it first): To go fast, go alone. To go far, go together.

Let's revisit the three questions:

- Do you serve the same user?
- Do you serve the same core use cases?
- Do you share the same buyer?

If the answers are not all "yes," you've found yourself a potential partner. The best partners are simply other companies' products that your customers use alongside your products.

Chapter 13 >_Enemies, Friends, and Frenemies...As If

Technology partnerships are *never* about getting another company to sell products for you. At best, you might have some joint events that add leads to the pipeline, but the rare cosell success of any technology partnership, one when the partner brings you leads, is a result of a solid field team that has built lasting industry relationships. You might have channel and systems integration partners (these often do consulting work for your customers) that sell for you, but if that's your primary goal of a technology partnership, it's going to fail. To make a technology partnership successful, you need to make a simple list:

- What do they have?
- What do they want?
- What do we have?
- What do we want?

The answers to these questions often include things like the following:

- Large user bases
- Brand reach
- Technology integrations
- Access to open source communities
- Enterprise customers

Only go after the partnership if they have something you want. Land the partnership by knowing what they need and showing how something you have can further their mission. Focus your energy on the narrow scope of mutual benefits.

Alas, there will be times when you encounter a company that has the same users, extends your user reach into new communities, has the same buyer, and even the same story. In times like these, you've either found a great acquisition target or, as we said earlier, competition that validates your market.

If you don't have competition yet, you will. These are often your frenemies. These days, an open source company hasn't crossed the chasm until a hyperscaler launches a competitive service. Taking your ball and going home isn't an option so you need a new plan.

Open source companies build software that is consumed by a community of users. The company makes money by selling services, support, or an enterprise version of the software or by delivering the SaaS version of the open source project. If your strategy includes selling your products and services to 5 percent of the community, a bigger community should mean a bigger selling opportunity. Bingo! Hyperscalers have broader reach and *way more* brand power than you. Work with them to accelerate open source adoption. As long as you can deliver products with differentiated value, the hyperscaler frenemy is an accelerator of success.

We've worked with startups that, before having a single customer, considered a hyperscaler to be their competition. Unless your startup was founded by a team of broligarchs, the hyperscaler cannot be your early competition (looking at you, OpenAI). You, so very literally, cannot compete. If you're seeking a Series A or B, your early pitch is that there's a change you seek in the world. Others want and need it, and you have a better chance of solving the problem and achieving the change than any other company. Competitors validate your market, and you pitch your differentiation or your unique point

Chapter 13 >_Enemies, Friends, and Frenemies...As If

of view. Save the "We're a hyperscale-esque platform" mantra for your Series C. Investors will eat it up at that stage.

As in life, friends compliment you in front of and behind your back. Partnerships require mutual ground. Hyperscalers are tools for growth, sometimes competition, and on occasion, potential investors. Competitors only compliment you behind your back in closed rooms when trying to figure out how you outsmarted and outstrategized them again. But none of these entities are enemies ... because they are not people.

> _Chapter 14

Get That Funny Money, Honey

Getting funding comes down to people, timing, and execution.

I was once at an investor conference where they organized pitches like speed dating. Every thirty minutes a new set of investors entered the room, and I took center stage. I gave the elevator pitch, and they asked hard questions about company metrics. At the end of the day, a banker I knew who worked for the conference organizers came in to talk. He explained to me that bankers are going to eat me up and spit me out for sport. He explained that this was not the place to wear my passion on my sleeve.

That evening, I wore a ten-year-old thrifted Prada dress to their dinner. They discussed how they race Porsches on the weekends and how they have hired help to get their backup Porsches to the next events. I realized what he saw. I was entering a new arena unarmed.

If you're going to get funding, you're going to need to get tough. Our cheerful, innocent demeanor in the rest of the book works well, except when the wolves are looking for the sheep. Your great idea is like their next Porsche; interesting, but replaceable.

The advice in this chapter looks like a hit list of things you must know because it is. This is not only our advice but also includes the advice of great founders and great investors.

Consider this your field guide to getting tough and navigating the hunt—where the future of your company is negotiated with guys that wear vests on purpose.

The People

Depending on your -vert status—extro or intro—the idea of pitching to VCs is either high-octane adrenaline pumping or high-octane adrenaline sucking. Either way, it's the ultimate test of your vision, your nerves, and your ability to sell not just a product, but a vision.

Cofounder Dynamics

Folks usually think cofounders are a balance of tech and business savvy. I spoke to a serial founder, Matthias Broecheler, who had a different perspective, and I love it. He looks for a cofounder that is unreasonable in the best way. They don't take no for an answer. They don't think inside the box. They'll do whatever it takes to do whatever it takes.

Know Your Leverage

Venture funding isn't charity. What is *your* leverage?

Are you a previous founder or industry influencer with a killer reputation and track record of success? If so, that's your leverage. Just send up a smoke signal and you should get funding quickly. Investors know that startups will change and pivot. Investing in experience is more reliable than believing in something or someone new. You

Chapter 14 >_Get That Funny Money, Honey

probably don't even need code, but you might need a cofounder that fills your personality gaps, which are very real and concerning to even your closest VC friends.

Not a seasoned founder? You'll either need a country club membership or code with traction. I'm not joking. A tech-savvy extrovert that's great at golf can network their way into a round of funding. I don't make the rules, I just report what the VCs told us.

The other 99 percent of us will need code with traction to get noticed. The capability should show unique value, and you should have traction that proves to investors that you're on to something.

VCs are drowning in potential deals yet starving for genuine gems. Your leverage is either your experience, your bullshitting/networking ability, or your code.

Educate Yourself on VC Motivations

In the early 2000s, before buying a used car, I ran a CARFAX report, found the previous owner, and called them from the parking lot of the dealership to ask about the history of the car. Fucked up? Sure, but why do I care? If I'm buying a car or taking funding from a VC, I want to know the truth about their mileage.

Not all VCs are created equal. Some seek early exits; others invest in long-term growth. Some are swayed by technological innovation, while others prioritize business models. Aligning with a VC whose goals mirror your own can be the difference between a fruitful partnership and frustrating micromanagement.

The first chapter of Adam Grant's book *Give and Take* discusses an entrepreneur initially choosing to exclude one of the greatest VCs out there, David Hornik, from his first investment round. David appealed to the founder's heart and passion, but another appealed to his need for business acumen above all else. The entrepreneur felt the need to trust his head over his heart and didn't choose David initially. The decision left him with regret, and he included David in the funding round before it closed. Why? Because heart is a main ingredient of any startup, and having an ally in your corner who is also one of the best is rare air. While the other investor bowed out early on in the venture, David and the founder continue to work together fifteen years later.[22] These aren't rash decisions. Spend time with your potential investors.

Prepare for Rejection

Securing VC capital is daunting. It involves numerous meetings, negotiations, and yes—rejections. Resilience is your greatest asset. Learn from each pitch and refine your approach.

Go Talk to Other Founders

I once went to a forty-five-minute conference session on VCs behaving badly, and my takeaway was that good founders want to pave the way for new founders. Just go talk to them. One guy left his hand up through others talking to ensure that he got the chance to say, "Side letters are not enforceable as part of your contract." The room erupted with stories of having learned that lesson.

Another person talked about how he took funding and the VC offered to throw a huge party at his mansion and assured the founders that every big industry player would be there. The VC delivered and the founders were elated, until they received a bill from the VC

Chapter 14 >_Get That Funny Money, Honey

for $80K. Others jumped in with stories of VCs expensing their Uber rides and meals after insisting on paying.

Timing

You must leverage industry cycles. Understand the ebb and flow of your industry's funding cycles. There are times when money floods into certain sectors (think AI, blockchain, etc.) and times when it recedes. Timing your funding rounds during these peaks increases not only the likelihood of getting funded but also the terms you might secure. This means it might be smart to take money before you need it. #AlwaysBeRaising

Execution

Clarity Is King

A common pitfall for many founders is a lack of clarity. Walking into a pitch with a muddled message is going to fail. Your value proposition should be crystal clear, and it should articulate exactly what your startup does, whom it serves, and why it matters.

Signal Strength

For new founders, your traction speaks louder than your projections. Traction can get you to the top of any VC's darlings list. *This* is why open source is your friend. You must shed your introversion to get your code seen. You're probably already part of the community. How else would you have found a problem worth solving or an idea worth pursuing? It's time to speak up. If this is hard for you, go read the transcript of or watch the YouTube video of Richard Hamming's talk titled, "You and Your Research."

Signals include user growth, engagement metrics (think GitHub stars), and technical validation. Are users sticking around? Are developers contributing to your platform? These signals help VCs justify the risk. You are the risk.

The Deal

Many founders think that they are in the weaker position for negotiations. Flip this on its head. If a VC needs your innovation to enhance their portfolio, you wield the power. Use this to negotiate terms that favor your long-term vision over immediate gains, preserve your equity, maintain control, and minimize board interference. You're here to make a mark on the world, not just to do a deal.

Seek terms that protect your vision. Opt for funding instruments like SAFE (Simple Agreement for Future Equity), which delay valuation discussions and simplify governance. Negotiate hard on clauses that impact your control and future funding rounds. The goal is to secure the money you need without selling your soul or giving up any control of the company that you don't have to. Pro tip: There are organizations that provide templates for supposedly founder-friendly term sheets, but these are written by VCs. They are not friendly. Get a lawyer, the best you can afford, and know if they also represent clients with competing interests to yours.

You can't avoid every pitfall in the funding process. Sometimes funding is really hard to get, and you have to take the deal in front of you even if it's a bad one. Eyes wide open, my friend. Don't tell yourself stories about how it's not that bad. Accept the situation for what it is.

>_Chapter 15

Piercing the Dark Veil of Acquisitions

I was at both public and private companies through their acquisitions. At my first, a public company, I and hundreds of others who had also just watched their dreams of being independently wealthy fall apart, waited patiently for new job offer letters from the acquirer. I'd also just received an order for seven hundred T-shirts that were supposed to be giveaways at an upcoming conference. Word came down from on high that we were not to give anything out with the old company logo. In the end, I got my offer letter and seven hundred T-shirts. I still have five or six of them, as does just about everyone close to me.

Years before that acquisition, I was sitting at the annual sales kickoff meeting in Utah. The evening had a Hawaiian theme, and there were tables stacked high with Tommy Bahama shirts as giveaways. Steve Wozniak was our chief scientist, and he kicked off the evening with a hilarious story about attempting to trick Nintendo into continuing to publish his top Tetris scores even though the magazine wanted to give others an opportunity at the top spot. In classic Steve form, he ended up tricking himself when he submitted his score under an alias, forgot, and then got mad when the alias was published with their/his top score. After that, the CEO got up and talked about how, through our hard work, we were all going to be independently wealthy.

I walked up to a friend after the talk and inquired, "What the fuck does 'independently wealthy' mean and why is it better than any other flavor of wealthy?" Neither of us knew the answer, and we reasoned that it probably meant that you had your own wealth and didn't need your parents' money. We were wrong, but it turns out it wasn't knowledge we needed anyway.

Woz kicked off the night with a story about accidentally beating himself at Tetris. The CEO followed with a speech about us all becoming independently wealthy. Only one of them was telling the truth.

I witnessed the cycle of invention, democratization, and commoditization with some of my closest friends and made a few more along the way. I got a lot from that experience, but as it goes for most startups, independent wealth—meaning that you have enduring financial freedom with the ability to support yourself—was only

Chapter 15 >_Piercing the Dark Veil of Acquisitions

realized by 1 to 5 percent of the employees. The rest of us got some harsh lessons about taxes and some great memories.

The acquisition itself intrigued me, though. It all felt so secretive and mysterious. I started paying attention to high-tech acquisitions like my friends paid attention to the Kardashians. I've admitted in a previous chapter that I thrive on industry knowledge, but this quickly became more than a guilty pleasure. I deeply wanted to understand the strategies, the motivations, and the goals behind these big moves, and I wanted to know years later whether they were the right strategies, motivations, and goals.

It would be more than a decade before I got to work on my first acquisition effort and a few more years before I got to run the process end-to-end with one of those dear friends I met while making memories at the other company.

What follows is how I think about acquisitions from a buyer's standpoint and how I think founders should look at them as well. This also warrants a disclosure: You're taking advice from someone that uses the *Mergers and Acquisitions for Dummies* book as a reference more often than I reference the authoritative books that I also own but find too wordy.

Why Do an Acquisition?

You can't buy a company strategy, but you can accelerate one through an acquisition. This is sage advice whether you are the buyer or the seller. If you're the buyer, you need stakeholder buy-in on what the goal of the acquisition is. If you're the seller, it's a dang good idea to decipher the buyer's strategy so that you can appeal to it and maximize your value.

Acquisition goals are usually some combination of the following:

Enhancing the product or filling a product gap: If your company is losing deal after deal to a competitor for a reason that you can mitigate through an acquisition faster than you can by throwing more developers at the problem, go for it. I see this as a good strategy when the acquiring company specializes in one thing (say, usability) and thus they don't have deep expertise in performance.

Moving into an adjacent market: The upside of entering a new market is that you have a new market to sell into. The downside of entering a new market organically is that it's hard to do. Buying code is often faster than writing code, and you can acquire the trust of the community and relevance within that new market. Buy something tiny if you have to, but remember that you're buying relevance in a market that is new to you.

Increasing your geographic reach: Investors don't like an overly concentrated customer base. If 90 percent of your business is US based, acquiring the European competitor will consolidate the market and excite your investors.

Consolidating the market: This is the classic "kill/embrace the competition" strategy that can reduce costs and competition while giving your company a boost as the perceived market leader. This is also where a lot of private equity firms put their focus, and leading the consolidation results in a better chance of you leading the combined companies.

Chapter 15 >_Piercing the Dark Veil of Acquisitions

Acquiring talent: A small consulting firm with broad industry trust is always a good acquisition. There are times when a good company just didn't get funding. The product might not be exactly aligned with your strategy, but the talent they have can still accelerate the strategy.

Playing the metrics game: A public company reports revenue and expenditures separately. All revenue from an acquisition is reported as new revenue to the acquirer. This can be a solid boost for the acquirer.

Other reasons: One company's operating losses might be another company's tax reduction. Refocusing a company is often best accompanied by an acquisition. There are a lot of times when a company has too much of one thing. It might be too much revenue from one or two customers and thus they need to diversify. It might be that the company has done really well in a market that is becoming more saturated.

Let's talk about how to unpack these high commandments of acquisition strategy.

Advice for Buyers

If growing your market means moving into adjacent tech areas, you must deeply understand the industry landscape of your market and the adjacent one. An industry landscape maps out the technology areas of that industry and which vendors and open source projects occupy each area. The landscape is more than a 2D grouping of company logos; it tells you how the different sectors interact, compete, and collaborate. You'll start to understand who needs who, what areas make sense to expand into, and which technology your customers must use to get the most value out of your products. Knowing this is the key to going from understanding the corporate strategy to writing an acquisition strategy.

VCs and open source foundations frequently publish industry landscapes. I assume that VCs need to know the landscape to be able to build a thesis for an associated fund, and publishing the knowledge is a step toward saying, "You want to get money from us because we know the industry and will guide you to success." (I'm an optimist, though.) Regardless, anything you can find is a good starting point, and unless something about each one annoys you, you don't know the industry well enough.

You only need to know one more thing before starting your search: affordability. Notice that I said "affordability" instead of "budget." You don't want to overly confine your searches. There are companies that have taken $5M in funding and will sell for $10M. There are companies that have taken $5M in funding and want $50M. There are also companies that have taken $20M in funding and are ready to sell for $15M (Obviously, you aren't going to pay the $15M).

Headcount is one of the biggest factors in whether something is actually affordable. Let's say each employee costs your company about $250,000 per year once you factor in salary, benefits, equipment, and overhead.

Over two years, that means:

$250,000 \times 2$ years = $500,000 per person.

```
Chapter 15 >_Piercing the Dark Veil of Acquisitions
```

Now compare two scenarios:

Hiring 25 people:	Hiring 5 people:
25 × $500,000 =	5 × $500,000 =
$12.5 million over two years	$2.5 million over two years

So here's the twist: If you raised **$10 million** and hired 25 people, you'll burn through it in about two years.

But if you raised **$20 million** and hired just 5 people, you'd have much more runway—and options.

Same total burn, wildly different team sizes.

The point: "Affordable" isn't just about what's in the bank. It's about how fast you're burning it, and who's holding the matches.

Make the List

It's time to make a list of acquisition targets! "Make a list" is too light a phrasing for this phase of the process. Although I rarely remember to take it with me, a list is something you make for the grocery store. We're not just "making a list," we're hunting! I don't hunt bananas at the grocery store. I don't have to. They're right there where I expect them to be. Great acquisition targets aren't sitting around on a shelf waiting to be picked up. They're often heads down being awesome and changing the world.

This also isn't a casual scroll through LinkedIn; this is a strategic operation. Use tools like Harmonic.ai, Crunchbase, PitchBook, or whatever database you trust. Set up specific queries, filter out the noise, and zero in on targets that align with your strategy. In summary:

- **Read industry newsletters:** Stay updated on who's making waves.
- **Analyze industry landscapes:** Identify companies that fit your criteria.
- **Monitor the competition:** See who the big players are partnering with or acquiring.
- **Leverage in-house knowledge:** Your team might have the next big lead.
- **Mine the databases:** Harmonic.ai, Crunchbase, and PitchBook are your friends (Harmonic.ai is probably your best friend).

Work the List

Narrow it down methodically. Keep notes on why you exclude certain companies—they might fit future criteria. Focus on factors that move the needle: user base, technology stack, and cultural fit.

This isn't just for your benefit. You'll need to present a compelling case to stakeholders who may not share your enthusiasm. Data and clear reasoning are your allies.

Chapter 15 >_Piercing the Dark Veil of Acquisitions

ADVICE
for Founders (Sellers)

Be findable: Make sure your company description is clear and straightforward. Ditch the buzzwords. Tell us what you do in plain language.

Be prepared: Have your financials, growth metrics, and vision well documented. Trusted advisors can be invaluable, but make sure they're truly in your corner. Advisors need to have more than just vested interests from your last funding round. Some investors are very solid advisors. Most are not. They often start with, "We won't sell for less than x." They aren't looking out for you.

Be open-minded: An acquisition offer isn't an insult to your entrepreneurial spirit. It's a potential opportunity. You don't have to accept, but consider the possibilities.

Sell before growth stops.

I recently talked to a founder whose company I acquired. His advice to founders is to establish trusted advisors early on. These are folks that can help you realistically look at a number of challenges and decisions you'll face while building the company. When his advisor showed up to our second meeting, I took it as a sign. The founders were open to the discussion but they had no intention of going into the process unarmed. The advisor you bring to the meeting should be unbiased to whatever extent is achievable. A board member from the VC that gave you money is *not* the person. If you're leading a tech company from Europe with Silicon Valley VC money, there's already a plan to force you to move or to replace you. VCs are business partners, not friends.

The second move the founder made was to get advice from a friend who had been through an acquisition. The friend's advice was to get the best M&A lawyer possible and trust that they'd be worth every penny. He did just that and I assure you, they were worth every penny. Nobody can read your mind. Actually, we (the buyers) can and we are, but if you don't negotiate it, you won't get it. I'm already making you rich. It's yours and your lawyer's job to make you richer.

The friend also advised him to keep discussions with his lawyers separate from discussions with his advisors. Advisors have a vested interest in the outcome of a deal. The lawyers were hired to serve the founders.

My advice? Be open to the idea!

Founders are daring and brave. An acquisition is just one more door that you can choose to walk through or walk past. It's true that most startups fail, but a great lesson from Y Combinator's Startup School is that founders rarely regret the experience. I continue to cheer for the founders of the many companies I've spoken to, and I'm lucky to work with founders that saw a better-together story with us. We're delivering on that shared vision, and both companies are more valuable as a result.

Chapter 15 >_Piercing the Dark Veil of Acquisitions

Let's All Talk

When someone reaches out for a "strategic conversation," take the meeting. As a founder, you'll learn why a bigger company thinks you're interesting, and it might not be what you expect.

As a buyer, when I show up to the first discussion, I know everything that is publicly available on both you and the company, and I'm trying to look well prepared without looking insane. If, however, you mention a tweet from ten years ago and see a flash of recognition on my face, it's genuine. It's my job to be curious about your journey. I want to talk because I think you're building something amazing, and I want to know what journey brought you to this moment. Uncovering red flags, getting to know your personality, and assessing your motivations are also my job, so I'm doing that too.

Founders are busy, and I don't expect them to do any work ahead of our conversation. But if *I* were the founder, I'd be curious. I'd want to know: Why me? How did I get on *your* list? How does what I'm building help you? And how the hell did you find my personal email address?

Better Together

When the prospect of an acquisition surfaces, it's all about how we can achieve more united than apart. If you don't see that potential, that's fine. Timing is everything.

But if our conversation ignites a spark, if we start envisioning a future where one plus one equals three, then we're onto something special. The right offer is one that pays ahead of a bunch of risk.

Keep in mind, offers often include a mix of cash and equity, with vesting schedules designed to keep you invested in the shared future. The real upside is tied directly to our combined success.

"Closing" Remarks

Closing a deal is not winning. Closing the right deal is winning. I've been so wrapped up in the excitement of a deal that I almost missed signals that it would have been the wrong move. As a buyer, be perceptive of the questions that are coming your way and whom they are coming from. If others are having trouble seeing the value, you might be more excited about the deal than the better-together possibilities.

As a founder, revisit the shared vision regularly. If it starts to fracture, address it head-on. Better to have tough conversations now than regrets later.

Lastly, tell no one. Deals fall apart for a lot of reasons. Don't be the reason.

Whether you're a buyer seeking to accelerate your company's strategy or a founder contemplating an offer, the biggest predictor of future success is to what degree both parties believe that the shared vision will propel the combined teams forward.

If we cross paths in the wild, know that we'll be talking because I think there's a chance that we could create something extraordinary together. It's either that, or I think you might have a great story for the next book.

>_Conclusion
Life's a Pivot

I still can't sleep the night before a launch. My mind still spins with "Did we consider this angle?" and "What about this opening?" I love to prepare for things, so much so that I find it hard to rest. Always have, still do.

But sometimes there is just that something you can't name that is different. There's the right amount of tension in your throat that signals, "I'm stoked for what's next." It wasn't there before, but it is now.

For me, it's the same feeling I had in the practice pool before my first national cut. My training just clicked, my energy was hardly containable, and everything felt *easy*. I was the second of four legs on that relay when we clocked an NCAA national standard. But after that race, I regretted that I wasn't the *anchor* of the relay and we didn't win the race. We had just joined a celebrated and small list of women who could say they achieved anything that warranted national qualification, but there I was wanting more.

It's strange to remember my college self so vividly yet feel like a completely different person now.

Maybe it was the athlete within that made me reach for Silicon Valley. Maybe it is because I like to compete in the biggest arenas.

Maybe it's because I like the rush of solving invisible math puzzles with the overclocked and unsung.

Or maybe it is the right fit for my blend of drive and quirk.

Whatever the reason, stepping into this arena meant that I had to take a long, hard look at the most irritating, overclocked misfit and learn that I was looking at myself. Growth meant learning how to genuinely envision an Analytical's perspective and slow down around an Amiable. Yes, I want to meet your puppy over Zoom!

I've always loved Drivers, though. My mom is one. Urgent decision-making and operations are our lifeblood. Stumbling upon a place where speed and bluntness are actually valued has been one of the most unexpectedly calming side quests I've completed in the arena. There's nothing more soothing than finding a place where my sprinter reflexes and energy are strategically valued . . . at times.

Just as I got comfortable flooring it, I realized the real challenge was learning when to pump the breaks. Unlearning the idea that my goal is to be the best, fastest, most prepared, and smartest person in the room has taxed my soul more than the restraint required to not breathe during that last lap. Working against your deepest and most ingrained behaviors requires the calmest, most centered focus I've ever conjured. It's very unnatural at first, but that's what growth feels like.

I've accepted that sometimes the real bottleneck wasn't bad code or clueless leadership. It was me, charging ahead at full tilt (Hulk smash!) while everyone else was just trying to finish their morning beverage.

And sometimes, self-awareness isn't just about realizing where you are the problem. It is also about resisting the urge to fix everything yourself.

Growth means holding others around me accountable for what I see.

Will They Say Yes?

Here's the crazy part: Letting go of my need to know it all didn't mean I abandoned my big ideas.

Quite the opposite.

The moment I stepped off my ego pedestal, I saw the market crying out for a solution built on scarcity and powered by relentless flywheel momentum. In other words, we've got a golden opportunity to launch a merry-go-round that you will want to push, especially if we move fast and strategically.

I think the stories, strategies, and conundrums we just shared in this book are just the beginning. I know there's more. And you have them. What do we, Denise and Kat, have in this partnership? Experience, heart, and a love for writing.

Our experience taught us how to see the abundance of secret plays, insider baseball tips, and mental breakdowns as opportunity. I cringe at "opportunity," but that doesn't change the reality that so many of us in this arena are struggling. Struggling to connect to one another and our bodies. We're just along for the ride, circling the merry-go-round with the apathy of a zombie.

We know how to fix merry-go-rounds. We know how to reconnect your flywheel to make it spin. We've faced enough competition to know who is friend or foe. We want to help you know these things, too. Reach out to us: We want to hear from you.

But remember, the path ahead starts with fixing yourself.

Here

You'll have the opportunity to repeatedly learn the lessons we shared in this book. I'll let the following story speak for itself. How many lessons can you spot that I forgot while having a moment?

Just before we started writing this book, I worked on a trajectory-changing acquisition. The trajectory change was very intentional. We needed to acquire a hot GenAI startup that fit our technical strategy and that was on the brink of mass adoption. We had great products, but we needed to buy a seat at the AI natives' table while paving a path from mass adoption to our high-value products.

I pitched the target founders on the potential of our combined company. They were fierce negotiators, and if a fraction of our shared vision became a reality, it would be life-changing for everyone.

We acquired the hot startup, and our trajectory did change.

Here I sit, ten months later, flying back home to Mississippi for a funeral, which is so f'ing ironically scheduled to be at the exact time we announce the death of our combined company via an acquisition by a tech giant.

The exit will be life-changing financially for those founders and our executive team . . . but less so for me.

I'm looking in the rearview mirror of ten years working toward an announcement that is now fourteen hours and ten minutes away.

I had been balancing writing and working on the big exit for nine months. The acquisition team has crushed over six hundred due diligence requests since Christmas, and the inertia of the effort is hitting a brick wall of reality. It's finally hitting me that all of this is all in the past and I have no idea what the future holds.

In 2014, I attended a new-hire training in California with my seven-month-old daughter, who is now an articulate eleven-year-old. A year into my tenure, my son was born, who is now nine years old.

I grew up here, made lifelong friends with the overclocked misfits here, learned here, broke myself here, found myself here, let go of my ego here, became a leader here, survived a pandemic here, became tougher here, and now *here* is gone.

My mom thinks I never cry. I've explained that as a woman in tech it's a learned talent. Tomorrow, I can give her the gift of emotion as I sit at a funeral listening to a sermon that is surely going to be about the need for a reckoning with one's own life before it's too late. I know this will be the topic of the sermon because Southern funerals are when the godly can witness to the heathens uninterrupted.

I get that tomorrow is not about me. I wrote the chapter on it, but I am facing a reckoning tomorrow. Ten years of my life for this . . .

In the coming months, I'll be offered a retention package while good people lose their jobs. The retention package is meant to make up for the "disappointing outcome." Not bad, but disappointing . . . I am disappointed.

One of the people I respect the most here, whose generous sharing of knowledge comprises many of the lessons in this book, will not miss this place. He'll be proud of what we did here, but he'll never miss it. Nothing has actually happened yet and I already miss it. I'm jealous of his framing, but it is not my framing. Damn I wish it was my framing.

I'm not sure if I'm more scared of who I'll become if I stay or of the unknown ahead of me if I leave it all behind. I want to use my superpowers for good. I want to inspire the next generation of startups.

There's one job out there that I've been noodling on for a year. I've written the outreach email in my head a hundred times.

I may send it. I may never send it.

Wherever I am when you read this book, know that I walked into the next opportunity with confidence, wit, grit, and a healthy appetite for growth. I expect setbacks. I'll be on the floor at some point wondering how I walked right into another trap. As Brené Brown says, those are the moments when we can look around to see all the other badasses that also dared greatly.[23] I hope to see you there.

We also hope that in those moments, this book is at eye level.

< one day later >

It turns out that the day was not about me. It was about a woman whose friends, siblings, children, grandchildren, and great-grandchildren gathered to honor her, and honor her we did. I was spot on about the sermon, but it was tastefully done and undeserving of my quip. It was, quite simply, a beautiful service.

There was a lot of fear, a lot of anxiety, in sharing the story of our company's transition. Sometimes you'll feel a duality in your framings of situations. What I felt was real. What is also real is that we turned that company around to be a desirable asset. We operated with a sense of urgency for the last five and a half years knowing that time was against us. I know that roles that are duplicated in an acquiring company are quickly eliminated. But I care deeply about our team and think that justifies a little pettiness in my emotions.

The truth is that I passed on many good opportunities to see this journey through. The experiences and opportunities for growth that I've had here are more valuable than the exit I had imagined. It's my job to decide what the next step will be. It's up to me to decide what my future holds.

Denise and I are both facing the unknown in this moment of our lives. We will walk into the next opportunity with confidence, wit, grit, and a healthy appetite for growth. We will expect setbacks. We will be on the floor at some point wondering how we walked right into another trap. In those moments when we look around to see all the other badasses that also dared greatly, we really really do hope to see you there.

We also genuinely hope that in those moments, this book is at eye level.

New-hire training, 2014.

>_Acknowledgments

To the Door Openers

To everyone we interviewed for this book. You know who you are, and we'll always honor your trust. You shared your stories with us when it was hard. You told the truth during pivotal moments in your own careers and let us learn from them. You helped us reframe our pain into something bigger.

Your honesty made this book. Thank you.

Ultimately, this book is for the door openers. The ones who showed up. The ones who reached out. The ones who reminded us that we don't do any of this alone.

Ever forward, together.

>_

Marrying Jon Erickson, the guy that sat across from me in a basement lab at the MITRE Corporation, was my great achievement in life. MITRE friends will chuckle at this, knowing I picked the seating arrangements in the lab. His love and support are my fuel. Our children, Karoline and Rex, are the reason I take my next breath.

To my parents, Jackie and Wayne, who, among many great parenting achievements, always nurtured my love of writing. On my eleventh birthday they gave me a homemade spiral-bound book whose pages I would fill with stories, and then many years later gave me a homemade quilt which was embolic of one of the poems from that book.

To Jenny Tindall, my big sister and life sherpa. Your support has given me great strength when I had none, great friendship, and endless great stories.

To my dear friends Mara Stuart, Chandra Ozkan, Cori Wolfland, Jackie Peltokangas, and that friend whose name I promised would not appear in the book (Matt Kennedy). To my cousins that keep me grounded and keep me laughing and to the BSB's and to everyone I am lucky enough to call a friend.

To my Hapkido family, painting crews, art friends, and fellow rockhounds. From each of these groups I have learned so much and found so much peace.

To the mentors and colleagues who shared their knowledge and provided the encouragement and feedback that I needed to grow: Christian Shrauder, Bill Hill, Ed Anuff, Jim Bergkamp, Sam Ramji, John Bangert, and JR West.

To all of my friends from MITRE, Fusion-IO, DataStax, Holmes, Southern Miss, and Hopkins. I am so grateful to have experienced such a range of follies, successes, and everything in between. The bonds we share comprise many of the stories of my life.

To everyone at Ideapress and TCG Worldwide who made this book a reality.

To Denise, whose friendship, dedication, and determination made "let's write a book" an actual book that we are both so very proud of.

And to my coach, Faith Halter, through whom I conjured the resilience and perspective needed to level up in life.

_Kat

Acknowledgments >_To The Door Openers

This book would not exist without the door openers in my life. The people who cracked something open when I couldn't even see the knob. Who helped me kick it down when I needed to, or who stood quietly behind me whispering go.

To Ty. My partner. My co-builder. My home. You have stood beside me through multiple books, but this one transformed us. It changed how we live and how we dream. Thank you for choosing growth. For putting in the work. For building something beautiful with me. I cannot wait to see what comes next.

To Kat. You saw me on the floor, literally and emotionally. You didn't flinch. You reached down, offered your hand, and helped me get back up. We found our voice together. We turned breakdowns into story arcs. We made something beautiful out of this book. I have never known a friendship or sisterhood like this. Thank you.

To my mom and dad. You were the first door openers. I grew up in the South in the '90s with a mom who worked in corporate logistics as the lead negotiator for international freight rates and a dad who stayed home to take care of us. That reversal was rare in our world, but you made it work with grace and grit. My mom taught me how to walk into a boardroom with purpose and strength. My dad taught me to question assumptions and to care deeply about people. Together you gave me permission to think differently about what women could do and how families could work. You didn't just support my path. You helped build it.

To my brother. Our competitive kind of love pushed me further than I thought I could go. You gave me the kind of challenge that sharpened me and made me better. That energy shaped me more than either of us probably realized.

To my family. You've had to ride this roller coaster with me. Through the triumphs and through the disasters. You've stood by while I tried to do really big things. You've celebrated with me, and you've held on when it got bumpy. I'm so thankful for your unconditional support and willingness to ride the highs and lows along with me.

Kat and I wrote this book during the year it took me to stand again after a serious, life-altering burnout, an injury I'm still recovering from. I would not have made it through without the support of a wide network of people who came to my side when I needed them most. To Brooklyn Milner: Thank you for helping me heal in ways I didn't know were possible. You brought me back from the deepest burnout of my life and helped me stitch myself back together in a way that made this book possible. You changed my life. To Dr. Mike Batla: Thank you for witnessing the depth of pain I carried while writing and recovering. You guided me through some of the hardest moments of my life. You gave me hope and became a friend exactly when I needed one.

To Elizabeth Buske and Michell Champ: you were my guides as I stepped out of the corporate world. You didn't just cheer me on, you asked the hard questions and stood beside me while I figured out who I was becoming. You helped me believe I could build something new, and I will always think of our weekly tribe as the anchor in my pivot.

Professionally, I've been just as fortunate. I've had the pleasure of navigating my career alongside some of the most generous door openers in tech. These are the people who didn't just make introductions; they made space. To Ted Tanner: You opened the door to

Acknowledgments >_To The Door Openers

Silicon Valley and helped me find a place in it. You've mentored me for years and shaped how I lead and what I build. To Dr. Matthias Broecheler: You built a vibrant open source community and took a chance on someone you barely knew. That chance turned into one of my longest and most trusted collaborations. You helped me learn how to find my way through startup chaos and in open source. I'm a better builder because of you. To Jonathan Lacefield: You mentored me into becoming a product thinker and hiked 50 miles on the Appalachian Trail with my crazy ass. You taught me how to ask better questions, both in the boardroom and when we're out of breath on uphill mile 7 post burgers and beer wondering why we thought this was fun. To Sam Ramji: You continue to open doors in all the right ways. You see the passion and the pain and still make room. You challenge me with care. You bring vision and generosity into every room you're in, and I'm grateful to be learning from you. To Brad Beebe: You opened the Amazon door and made space for me to learn from the best. When I hit rock bottom, you stayed in my corner. Thank you for supporting me when I needed it most.

Bringing this book to life meant finding my voice all over again. Sharing in public, especially about the messy, raw parts of your career, is a special kind of vulnerability, and I couldn't have done it alone. I was lucky to be surrounded by a publishing team that didn't just guide the process but helped me rebuild from the ground up. To Bryan Wish, Rohit Bhargava, and Charlie Fusco: You helped me build my next chapter from scratch. You've seen the chaos. You've walked through the identity shift with me. You never blinked. Your support has meant more than I can say. To Eric Koester: You built a writing course that actually works and taught me how to take stories out of my head and make them land. This book wouldn't exist without the structure and systems you taught me. You introduced us to Trisha Giramma, whose early feedback cracked the story wide open when the emotions felt too heavy to carry alone. Trisha, thank you for helping us turn emotional blocks into scenes on the page and for keeping the momentum alive. I'm so grateful for that. To Herb Schaffner, the best editor I have ever worked with: You helped us nail the final cut when the deadline clock was ticking and the structure felt impossible to land. You brought clarity to the chaos and made the ending sing.

We are never just the work we produce; we are also the community that carries us. I've been lucky to have incredible communities behind me for decades, people who saw me through each chapter and helped shape who I became. To every parent who trusted me to coach their kid in swimming when I was still a kid myself, thank you. That was my first try at leadership. I made every mistake imaginable, but I also found my style. Those years gave me a foundation I didn't even know I was building. To my friends from tennis to swimming to the Appalachian Trail, you've known me through all my phases. You've seen the overachiever, the neurodivergent spaz, the meltdown, and the rebuild. You didn't just stay; you cheered. I will never take that kind of friendship for granted. Special thanks to Dr. Katherine Schumann: You stood with me through the darkest parts of my PhD. You made sure I made it through, and you are still a collaborator and someone I trust deeply. To Dr. Michael Berry: You gave me a hand when I didn't think I could make it. You've become a lifelong friend and a compass for what it means to show up for someone. You are part of the reason I finished. To Dr. Teresa Haynes: Meeting you changed everything.

Acknowledgments >_To The Door Openers

You didn't just teach me math; you taught me that the way I see the world actually makes sense. You gave me a home in graph theory and AI, and a place in a field where I could thrive. I have spent my whole career walking through the door you opened.

To all of the overclocked misfits I've worked with. Thank you for shipping with duct tape and dreams. For staying late. For triaging on the fly. For solving problems no one else wanted to touch. You made the impossible happen again and again with little to no credit. Your grit made the work better. I see you, and I'm grateful.

_Denise

WORK WITH US

If one of the stories in this book reminded you of your own and you're ready to share it—or if you're curious about our upcoming workshops—we'd love to hear from you. Reach out anytime at **hello@techconfidential.ai.**

Let's keep building the kind of tech culture we actually want to work in.

>_Endnotes

1. Holiday, Ryan. *Ego Is the Enemy*. Penguin Random House. 2016.
2. Walsh, Bill, Steve Jamison, and Craig Walsh. *The Score Takes Care of Itself: My Philosophy of Leadership*. Portfolio. 2009.
3. Holiday, Ryan. "The Important Thing Is to Not Be Afraid." *Ryan Holiday*: Meditations on Life and Strategy (blog). RyanHoliday.net. April 6, 2020. https://ryanholiday.net/the-important-thing-is-to-not-be-afraid/.
4. Arnsten, Amy, Carolyn Mazure, and Rajita Sinha. "Everyday Stress Can Shut Down the Brain's Chief Command Center." *Scientific American* (April 2012). https://www.scientificamerican.com/article/this-is-your-brain-in-meltdown/.
5. Garmus, Bonnie. *Lessons in Chemistry*. Doubleday. 2022.
6. Barry, Harry. *Emotional Resilience: How to Safeguard Your Mental Health*. Orion. 2018.
7. Vallerand, Robert J. *The Psychology of Passion: A Realistic Model*. Oxford University Press. 2015.
8. Costa, Rui. "In Pursuit of Pleasure, Brain Learns to Hit the Repeat Button." Columbia Zuckerman Institute. March 1, 2018. https://zuckermaninstitute.columbia.edu/pursuit-pleasure-brain-learns-hit-repeat-button#:~:text=The%20mice%20quickly%20learned%20which,the%20pleasure%20hit%20of%20dopamine.

9. Torre, Pablo. "Why the NFL Has a Jargon Fetish | PTFO." YouTube video, 42:38. September 5, 2024. https://www.pablo.show/p/why-you-cant-understand-the-nfl.

10. Wooden, John. *Wooden: A Lifetime of Observations and Reflections On and Off the Court.* Contemporary Books. 1997.

11. Might, Matt. "The Illustrated Guide to a Ph.D." 2010. https://matt.might.net/articles/phd-school-in-pictures/.

12. Might, Matt. "Being an Accidental Pioneer in Precision Medicine | Matt Might | TEDxBirmingham." TEDx Talks. May 2019. Video, 14:57. https://www.ted.com/talks/matt_might_being_an_accidental_pioneer_in_precision_medicine.

13. Might, Matt. "The Illustrated Guide to a Ph.D." 2010. https://matt.might.net/articles/phd-school-in-pictures/.

14. Freudenberger, Herbert J., and Geraldine Richelson. *Burn-out: The High Cost of High Achievement.* Anchor Press. 1980.

15. Van der Kolk, Bessel. *The Body Keeps the Score: Brain, Mind, and Body in the Healing of Trauma.* Penguin Books. 2014.

16. Stone, Brad. *The Everything Store: Jeff Bezos and the Age of Amazon.* Transworld. 2013.

17. *Deadpool & Wolverine.* Directed by Shawn Levy. Marvel Studios, 2024.

18. Musk, Elon (@elonmusk). "Franz throws steel ball at Cybertruck window right before launch. Guess we have some improvements to make before production haha." Twitter (now X), November 22, 2019. twitter.com/elonmusk/status/1197675169155676160.

19. "Tesla's New Cybertruck Passes the Glass Test This Time." *Bloomberg.* December 1, 2023. www.bloomberg.com/news/videos/2023-12-01/tesla-s-new-cybertruck-passes-the-glass-test-this-time-video.

Endnotes

20 Apple Newsroom. "Letter from Apple Regarding iPhone 4." Apple. July 2, 2010. https://www.apple.com/newsroom/2010/07/02Letter-from-Apple-Regarding-iPhone-4/.

21 "Samsung's 'The Wall' TV: An In-Depth Look at the Futuristic Modular Display." TechRadar. January 9, 2018. www.techradar.com/news/samsungs-the-wall-tv-an-in-depth-look-at-the-futuristic-modular-display.

22 Grant, Adam. *Give and Take: A Revolutionary Approach to Success.* Penguin Publishing Group. 2013.

23 Brown, Brené. *Daring Greatly: How the Courage to Be Vulnerable Transforms the Way We Live, Love, Parent, and Lead.* Avery. 2012.

>_Index

A

Academics 73, 74
acquisition 173–175, 188–189
    advice for buyers 177–179
    closing deal 184
    goals of 176–177
    make targets 179–182
    "strategic conversation" 183
admins 98–99
Adobe 133, 135
advice
    for buyers 177–179
    for founders 181–182
advisors 181–182
    modes of 75
affordability 178–179
allergies 60
Amazon 114
    growth strategy 113
    leadership principle 20
Amiables 91–93
amygdala 24
Analyticals 86–88, 131
anxiety 40–42
APIs 38, 124
Arnsten, Amy
    "Everyday Stress Can Shut Down the Brain's Chief Command Center" 23–24

B

Ballmer, Steve 17
Barry, Harry 28–29, 32
    Emotional Resilience 28
behavioral changes 59
Bezos, Jeff 113
The Body Keeps the Score (Van der Kolk) 62
Bourdain, Anthony 6
Broecheler, Matthias 168
Brown, Alton 6
burnout 55–57, 61–62
    stages of 57–61
buyers, advice for 177–179

C

C-level title 54, 56
code freeze 135
cofounders 168, 169
Collins, Jim
    Good to Great 111, 113
communication, cross-team 97, 98
competition 161–164
    monitoring of 180
competitors 161, 176
    market validation 164–165
    tracking and comparing metrics of 161–162
conflicts, displacement of 58
Creative Cloud 133, 135
cross-team communication 97, 98
Crunchbase 179, 180
curiosity 43–46
customer service 73

D

data
    drumbeat creation 150–151
    grounding of shared narrative 148–149
    hoarding/siloing 143
    trust and value in 124
    types of 153–155
    worse before better 152–153
data-being, states of 86
data scientist 95–96
deadline heroics 149–150
depersonalization 60
depression 61
Drivers 83–85, 88
Dunning-Kruger effect 46–47

E

ego 15, 17–20, 23, 25, 31, 51, 96–97, 149, 187
Ego Is the Enemy (Holiday) 15
emotional awkwardness 121
Emotional Resilience (Barry) 28
employees, costs of 178–179
employee stock program 99
emptiness 60
"Everyday Stress Can Shut Down the Brain's Chief Command Center" (Arnsten, Mazure, and Sinha) 23–24
The Everything Store: Jeff Bezos and the Age of Amazon (Stone) 113
experimentation framework 144
Expressives 88–91
extrovert, tech-savvy 169

F

FAANG companies 31, 33, 46–48, 59, 61, 64
failures 127, 136
    internalize of 28
fear and anxiety 24
finance 99–100
Financial Services Cloud 134
Firshman, Ben 108
Flagship Charts 134
flywheels 111, 140
    momentum 111–112
    self-reinforcing loop 112–116
founders
    advice for 181–182
    revisit shared vision 184
    "strategic conversation" 183
Freudenberger, Herbert 57
funding 77–78, 167, 171
    instruments 172

G

Gates, Bill 26
Give and Take (Grant) 169
Good to Great (Collins) 111, 113
gossip 159–161
Grant, Adam
    Give and Take 169

H

Hacker News 75
Hadfield, Chris 23
Hamming, Richard 171
Harmonic.ai 179, 180
high-tech acquisitions 175
Holiday, Ryan 23
    Ego Is the Enemy 15
"hormone darts" 25
Hornik, David 169–170
Human Resources (HR) 102
hyperscalers 101, 164
hypothalamus 24

I

imposter syndrome 40–42, 46–47
independently wealthy 173, 174
intelligence 159, 160
introverts 71, 168
investors 167, 176
"it's not about you" rule 29–33

J

job
authority required to be successful 54
resources needed to be successful 54–55
    scope of 54
    wanting of 53
Jobs, Steve 26

K

Kearns, Abby 29, 43–45, 48, 49
"kill/embrace the competition" strategy 176
knowledge 42, 44, 75, 178
Kolk, Bessel van der 66
    The Body Keeps the Score 62

L

leadership 75, 77, 96
    ask *vs.* tell 82
    meetings 147, 148
    open *vs.* closed 82
Lean In (Sandberg) 75
legal 101–102
leverage 168–169, 180

Index

M

marketing 95–96, 132, 147
Mazure, Carolyn
    "Everyday Stress Can Shut Down the Brain's Chief Command Center" 23–24
McConaughey, Matthew 38
Mergers and Acquisitions for Dummies 175
Microsoft 32, 63
middle management 8
Might, Matt 39, 40
minimum viable product (MVP) 136
momentum, flywheels 111–112
Musk, Elon 127

N

Nadella, Satya 32
Nuhook 120, 121

O

obsessive passion 33
open source companies 164
open source foundation 43, 178
open source licenses 101
operational data 153
ops team 96–98, 135
    set up, product launch 123–125
Orlovsky, Dan 37
outside-in thinking 130
overclocked misfits 71–79

P

Pang, Alex Soojung-Kim
    Rest: Why You Get More Done When You Work Less 91
passion 33–35
Patrick (Microsoft veteran) 17–23
Peter (VC) 16–17, 19–20
PitchBook 179, 180
The Practitioner's Guide to Graph Data 6
prefrontal cortex 24
pre-IPO startup 56
product launch 117–118, 127, 139
    countdown to 125
    design and build new ride 121–123

educated wishes 118–121
    set up ops 123–125
    vitamins *vs.* painkillers 129
product-led growth 139, 140
products
    growth of 143, 144
    launch day 136–137
    market pull 141
    naming 132–135
    positioning of 128
    pricing of 130–132
Publiccomps.com 161

R

R&D, production issues with 74
repeatability 139–141
Rest: Why You Get More Done When You Work Less (Pang) 91

S

SAFE. *See* Simple Agreement for Future Equity (SAFE)
Sandberg, Sheryl 2
    Lean In 75
scarcity–abundance S-curve 107–110
S-curves 107–110, 115
self-care 58
self-reinforcing loop, flywheels 112–116
self-worth 29
Sinegal, Jim 113
servitude 81
Silicon Valley 3–5, 7, 9, 45, 185
Simple Agreement for Future Equity (SAFE) 172
Sinha, Rajita
"Everyday Stress Can Shut Down the Brain's Chief Command Center" 23–24
Social Styles Matrix 82–83
    Amiables 91–93
    Analyticals 86–88
    Drivers 83–85
    Expressives 88–91
social withdrawal 59
software engineering 72
Spider Y 2 BANANA play 37, 40, 43

startups 56
    GenAI 188
    legal 101
    Nuhook 120, 121
    product launch 117
    sales numbers 149
Stone, Brad
    The Everything Store: Jeff Bezos and the Age of Amazon 113
strategic data 153
"strategic technical advisor" 74
stress 22, 24, 57, 62
SurveyMonkey 133

T

Talmudic expression 29
technology partnerships 163
tech-savvy extrovert 169
tech utopia 4, 9
TED Talks 81
telemetry 131, 151
Tice, Nate 37
Torre, Pablo 37
traction 171
    code with 169
trust 147, 148, 177

U

user journeys 141–143

V

Vallerand, Robert 33
values, revision of 58
venting 160
venture capitalists (VCs) 168, 169, 182
    educate yourself on motivations 169–170
    execution 171–172
    networking event 16
    and open source foundations 178
    prepare for rejection 170
    talk to other founders 170–171
    timing 171

W

Walsh, Bill 18
warning signs 51–52
"work-induced PTSD" 61
Wozniak, Steve 173–174

Y

Y Combinator's Startup School 107, 113, 141, 182